To my dear friend Donna,

You are pure awareness!

Thomas

Mystical 2012

The Triple Rebirth of the Sun and the Triple Rebirth of You

by Thomas Razzeto

ISBN: 978-0-9826563-0-3

Front and back cover design: Thomas Razzeto
Cover photograph: Thomas Razzeto
Diagrams: Thomas Razzeto
Screenshots: Thomas Razzeto with the astronomy software
program: "The Sky" version 6

Dedicated to my mother and father,

from whom the love of God

truly flows freely to me

Table of Contents

(continued)

Is this book for you? Find out quickly and easily by reading my free essays at 2012essays.com.

Preface

I have often said, "The more I dug into 2012, the more mind-blowing it all became!" Here's the story of how it unfolded for me.

In early 2008, I attended a lecture about science and myth. "2012" was not in the title but the lecture mentioned the Maya calendar and the astronomy of 2012. I became intrigued enough to do some research and I quickly found the excellent work of John Major Jenkins. I learned about the "galactic alignment" that had already happened in 1998 and how the slow, steady wobble of the earth brought about the astronomical conditions that the calendar and the Maya folklore seem to be talking about.

My initial motivation to write about this subject was to quell the fear of "doom and gloom" that is so common in the popular media but I still had some important questions. Why did the Maya pick 2012 and not 1998? Did they just miss by 14 years? My intuition said no, they picked 2012 for a solid astronomical reason. But what was it? And how did they know all this amazing astronomy? And of course my biggest question was this: Do the Maya have something important to tell us today?

I already knew that ancient people often saw the daily sunrise and the winter solstice as two different types of solar rebirths. Did the ancient Maya see yet another type of solar rebirth in 2012? As I dug deeper into the subject, that thought was in the back of my mind.

I began to focus on the disk of the sun crossing the galactic equator, as seen from the Maya's viewpoint on earth. This crossing will take about 15 hours and it will fit completely on the calendar date of the winter solstice in 2012. Could this be the sun being metaphorically reborn as it moved from one side of the galactic equator to the other? If so, this would create what I called "the sacred triple rebirth of the sun." This was my first hypothesis and I continued to look for more supporting evidence.

I later abandoned the idea that the Maya were concerned at all with the galactic equator, since it is completely invisible. I soon realized that the astronomy that is relevant to 2012 was astronomy that the ancient Maya could see with their naked eye. As amazing as it sounds, the Maya created this calendar without the aid of telescopes, computers or clocks! And in this book, I will show you why and how they did this.

While I could not know for sure if I was the only one thinking about 2012 as the triple rebirth of the sun, I can say that I did my own thinking without reading about the triple rebirth elsewhere. Although John's galactic alignment thesis was an important step in the development of my understanding, it emphasizes a significantly different aspect of the astronomy.

By the spring of 2008, my bubbling curiosity started to be transformed into awe. The triple rebirth appeared to be an excellent reason that the Maya picked 2012 rather than 1998. At this time, I did the calculations that determined that in order for the Maya to know the exact number of days until the winter solstice of 2012, they would have needed to know the length of the year to within 45 seconds! Now I was simply astonished. I found myself repeating, "It's a triple rebirth of the sun and the Maya needed to know the length of the year to within 45 seconds!" From this moment on, I never lost my sense of awe. I was continually amazed as I worked on more research and more writing.

Early in the summer of 2008, I posted a detailed astronomy essay on my website and I took two months to create a YouTube video. I even talked to George Noory during open lines and chatted with him for about three minutes. I did my first radio interview about this subject in August of 2008 with Richard Syrett in Toronto. The interview was titled: "The

Actual Astronomy of 2012 and the Sacred Triple Rebirth of the Sun."

In the fall of 2008, I started writing an essay titled "Mystical 2012." In that essay, which eventually grew into this book, I offered my opinion that the rebirth metaphor of the Maya was mystical in nature. I had not yet dug into the psychedelic rituals of the Maya shamans but it seemed to me that the rebirth metaphor pointed to a spiritual rebirth that could happen at any time and that this was so important to the Maya that it was at the core of 2012. "Mystical 2012" blended the astronomy and mysticism together in a way that I felt was both beautiful and relevant to the Maya and over the next few months, I posted several more 2012 essays on my website.

I wanted an astronomer to double-check my work since I had made some estimates based on the data from the Belgian astronomer, Jean Meeus. Steve Kates, also known as Dr. Sky, said that he did not have time to confirm all my data but he suggested that I get an astronomy program so that I could examine the astronomy myself without extrapolating from the conditions of 1998 offered by Meeus. I did so and spent a tremendous amount of time looking at the astronomy in every way that I could possibly imagine. Steve liked my work and we did a radio interview together in April of 2009. I gave my first public talk just the month before and it was titled: "Mystical 2012."

The triple rebirth of the sun was holding up very well to all the data and yet I still wanted to find a more definite link between the astronomy and the calendar. I wanted to find something that was obviously of interest to the Maya. John Major Jenkins encouraged me in my work and was very generous with his time by sending me a few e-mails in June 2009 clarifying his work and the work of others. One of the things that I wanted to clarify was this: In the years around 2012, there are many years that offer a triple rebirth of the sun. Why did the Maya pick the triple rebirth of 2012 rather than any of those other years?

During the summer of 2009, my focus changed from the galactic equator to the dark rift. These two things are very near each other but one is visible to the naked eye and the other is completely invisible, even with a telescope! This is a very important point! The Maya created metaphorical stories about what they saw with their eyes and they metaphorically saw the dark rift as the birth canal of the Sacred Mother. This made the triple rebirth metaphor directly relevant to the Maya since the sun will appear to be in the birth canal on the day the calendar restarts.

In July of 2009, I took one more very important step. I finally took a look at a much bigger section of the sky. Instead of keeping a tight focus on the sun and the galactic equator, as seen from earth, I opened up my field of view to include the planets around the sun on that special day. This was when I found what I was looking for: the Maya's "sacred tree" in the sky above them in a stunning configuration! I almost fell out of my chair when I first saw the image on my computer screen! (This screenshot is on the back cover. Please note that other people were already aware of this configuration so I did not make an original discovery; I simply found it for myself.)

I was now sure beyond a reasonable doubt that the combination of the triple rebirth of the sun and this unique configuration of the sacred tree was indeed the reason that the Maya had picked the exact day of the winter solstice of 2012. As of February 2012, I still appear to be the only person presenting this combination as the reason the Maya restart their calendar on that exact day.

But understanding the link between the calendar and the astronomy does not answer the biggest question: What are the Maya trying to tell us? While I had been thinking that the core message of 2012 was mystical in nature for about 18 months, in January of 2010, I began to look for solid evidence. Since I wanted to make sure that my investigation of the spiritual dimension of the Maya culture was well grounded in relevant facts, I started to study the shamanic rituals of the Maya.

This was yet again another stunning moment in my research. I learned that the Bufo toad was the source of a psychedelic chemical used by the Maya shamans and that there were stone altars carved in the shape of these toads along with other stone carvings prominently displaying this toad. This was strong evidence that in the eyes of the Maya, these psychedelic experiences were very important!

When I looked into the effects of this psychedelic chemical, I became convinced that long ago, the ancient Maya used it to discover the greater reality that lies behind our ordinary reality, and most importantly, they discovered their true fundamental self. When we re-identify with our true self, we are reborn in the most profound way. It made perfect sense to me that this spiritual awakening was the rebirth that is at the core of 2012.

I should also point out that starting in about 2006, I began exploring mysticism in a deep and personal way. I had been following a spiritual path for several decades but it was at this time that I was clearly presented with the deepest teachings of this tradition. I had been writing spiritual essays since 2001 but now my writing jumped up to a whole new level. So the seeds of *Mystical 2012* were planted a year or so before I began to explore 2012.

The material in this book is quite deep. I often went on hikes by myself to places full of trees, flowing water and large boulders in order to ponder many of the topics that I discuss here. If you can do likewise, I would encourage it. Some people might initially think that they can quickly read my book since it is so short but I don't think that that is the best way to really get the most out of it. Even today when I re-read it myself – which I've done many times – it still grabs me and makes me stop reading in order for me to yet again deeply ponder certain points. While this book is certainly not for everyone, you will know that it's for you if you find yourself doing just that!

By the way, I want to be very clear about something:

My work is a synthesis; it is the collection of previously known parts put together with a new understanding. I did not discover the sun in the dark rift; I did not discover the special configuration of the sacred tree. Nor did I discover the stone altars in the shape of the Bufo toad. Yet I put all of these things together in a way that uniquely explains why the ancient Maya created their calendar and what the message of 2012 actually means.

While I drew upon the work of many people, it is the work of John Major Jenkins that I would like to single out as the most important for me. John's work brought so many important things together in one place that my work was much easier. To be sure, John and I have our differences, but he is a without a doubt a great guy, a fine researcher and an excellent scholar. Many thanks to you, John!

Even though the most important ideas that I present about mystical spirituality are thousands of years old, perhaps you will find my presentation refreshing and easy to understand. The goal of this book is to offer the guidance that will facilitate a fundamental shift in the way you see both the world and yourself. People often talk about enlightenment and a sudden spiritual awakening. Can a book actually help make this kind of thing happen? Well, I think that it can and *Mystical 2012* is my best effort in that regard.

And to go further with that idea, please look again at the cover of this book. Do you want to be walking on that beach enjoying that sunrise? Well, this book is really about you, not 2012, and I want you to imagine walking down that beach, right *through* the sunrise, and right into the heart of this book. I want you to become so immersed with this book that you are transformed by the fire of the sun in ways you never before even imagined! So please take the time to carefully consider what I have to offer. Let these ideas consume you as they consumed me!

All my best in your adventure of self-discovery!

Thomas Razzeto
February 28, 2012

Overview

I like to ask people the following question:

***Do you want the message of 2012 to inform you or transform
you?***

Please keep that question in the back of your mind as we
explore all the ideas in this book. Also, please check all
preconceived notions at the door. That's right. If you have heard
other ideas about 2012, either about the astronomy or what the
message is supposed to be, please put those ideas aside for a
while. You can certainly pick them up again later, if you wish.
With that, let's get started.

Just over 2,000 years ago, the Maya started to carve dates from
their Long Count calendar into stone. While this calendar has an
overall length of 5,125 years, the Maya started to use it not from
its apparent beginning but from a point that set its restart date to
the exact day of the winter solstice in 2012. Why? What did
they foresee on that precise day so far in their future?

The first clue that astronomy is at the core of the calendar is the
fact that the restart date is a winter solstice. This is, of course,
the shortest day of the year and an astronomical event. This
inspired me to dig deeper into the actual astronomy of 2012 and
in chapter two, you will learn how this winter solstice is part of
an astronomical event that the Maya metaphorically saw as the
sacred triple rebirth of the sun.

Don't worry about the astronomy being too hard to understand;
this is more like art than astronomy! You can do this! The most
important points are revealed in a picture of what will be in the
sky over the Maya *and over all of us* on that special day.

Here it is (and it is on the back cover in color):

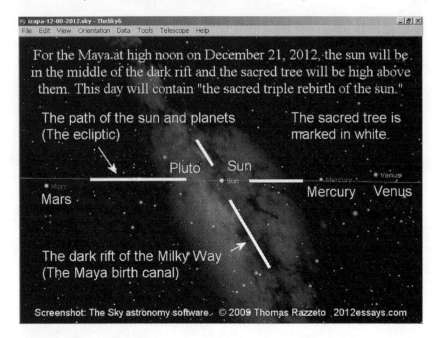

For the Maya, at high noon on December 21, 2012, the sun will be in the middle of the dark rift and the sacred tree will be high above them. This day will contain "the sacred triple rebirth of the sun."

The path of the sun and planets (The ecliptic)

The sacred tree is marked in white.

Pluto Sun

Mars

Mercury Venus

The dark rift of the Milky Way (The Maya birth canal)

Screenshot: The Sky astronomy software. © 2009 Thomas Razzeto 2012essays.com

What a beautiful image! See why I say this is more like art than astronomy? After you see how I present the astronomy of 2012, you might wonder how so many people can be completely unaware of this simple but important perspective.

It will be very easy for you to quickly understand why this astronomy is so special and why it was used as the foundation of the Maya's transformation and rebirth metaphor, which just might be the ancient Maya's most important metaphor of all! This is why I want to encourage you to try your best to read about the basic astronomy of 2012 in chapter two. Yet if you just want to glance at the pictures, that is perfectly okay, too. (On the other hand, chapter three is for people who really want to dig into the astronomy; it is perfectly fine for you to skip that chapter completely or just glance at the pictures.)

But what is the core meaning of this rebirth metaphor? This is the focus of the chapter four. What I offer in this regard is, of course, just my opinion, yet I have worked very hard full-time for several years to formulate an opinion that is based on facts that come from the Maya culture itself, not on misconceptions and over simplifications that can be found in the popular media. And even though it took hard work to formulate my opinion, what I have come up with is very simple.

According to John Major Jenkins, the purest meaning of 2012 is found by examining the stone monuments and carvings in Izapa, Mexico, the birthplace of the Long Count calendar. This is the most immediate and primary evidence left by the Maya.

Some of these stone carvings clearly point to the astronomy of the triple rebirth of the sun yet much more important than that are the stone altars carved in the shape of the Bufo toad and other stone carvings depicting this toad. It is from the Bufo toad that the ancient Maya shamans got the psychedelic chemical that they used to explore "the underworld." These psychedelic experiences are what allowed them to see the greater reality that lies behind our ordinary reality. This is what turns "2012" into "mystical 2012!" Again, after you see what I present in this regard, you will wonder how so many people can be completely unaware of this simple but important perspective.

It is my opinion that the Maya's timeless metaphor of transformation and rebirth is for anyone who wants to dive deeply into the true nature of reality, and that at its core, it points to a mystical view of reality and to a completely new understanding of the fundamental self.

I conclude the book with a chaper about mystical spirituality in general. While this might not seem directly related to 2012, I hope that you will still find it enlightening.

You now see that my approach to unlocking the mystery of 2012 is to think like a Maya skywatcher and to understand our world and the mystery of life like a Maya shaman. While I do not claim that I have been able to do either of these things to a high degree, that is what I am striving for and this book is the result of my best effort to do so. To put it simply, my approach to the puzzle of 2012 is to carefully consider its three tightly related aspects:

1) the astronomy that will unfold exactly on December 21, 2012
2) the Maya rebirth metaphor that is based on this astronomy
3) and the core meaning of this metaphor.

I hope I have inspired you to take a fresh, new look at this intriguing subject with an open mind.

But be forewarned: this material completely blew my mind! Maybe you will have a similar experience!

Chapter 1 - Introduction

Perhaps the truth about why the Maya picked the winter solstice of 2012 is simple. And perhaps the reason the truth is overlooked is because it is not as spectacular as a huge natural disaster or a sudden upward shift in human consciousness or any other dramatic scenario you may have heard about. Since no one can prove that these things will not happen, many people will hold on to these dramatic possibilities and the media will push them. But in my opinion, this focus on dramatic possibilities will blind people's eyes and prevent them from seeing the truth.

In my opinion, the reason the Maya picked the exact date of December 21, 2012 to restart their calendar is because of the astronomy that will unfold in the sky over the Maya *and over all of us* all on that precise day.

Almost all calendars are based on astronomy and it appears that this is also true for the Maya calendar, although in a rather spectacular way! And it is important to point out right up front that this astronomy presents no danger to us at all. We are not going to fall into a black hole or be inundated with a high level of cosmic rays.

In my view, this rare astronomical event will not cause anything unusual to happen to us at all, either good or bad, physically or spiritually. And yet, it is my opinion that the Maya's timeless message of transformation and rebirth can potentially leads us onto an exceptionally rewarding path of self-discovery.

Before we get off on the wrong foot, please let me clarify something. Of course I am well aware that many people think that 2012 is about a shift into a new golden age of world peace. Yet after years of research, it does not seem to me that the ancient Maya themselves created the calendar to predict world peace at this time. *But this does not mean that we can't create a beautiful world right now!* Is this just around the corner? I simply don't know. But I do know that that is what I am working for and that no matter what the conditions of the world might be, we all can make our world a better place by just naturally being ourselves and sharing the gifts the divine has given us all.

And still, I would like to take a few minutes to explore the possibility that 2012 is a psychic prediction. I think it is very significant that the calendar points to one specific day, not to a longer period of time. If the Maya foresaw something important that was going to unfold over many decades, for example, then I think they would have made a calendar that would obviously point to a longer period of time. Yet the calendar points to one specific day. When you understand the astronomy that will unfold on that day – *and only on that day* – and when you see that this astronomy is an excellent fit for the Maya folklore, then perhaps you will agree with me that the calendar restarts on that special day because of this remarkable astronomy.

And still many people think that the calendar is based on a prediction about something dramatic that will unfold in the time around 2012, and perhaps this event will be global in its scope. Some people think that doom and gloom is just around the corner while other people think that we are about to experience a sudden upward shift in human consciousness – or perhaps a combination of both!

But why would the Maya even try to look 2,000 years into their future? What would have motivated them to do so? What could have attracted their attention? If the ancient Maya were interested in events far in their future, wouldn't their own conquest at the hands of the Spanish invaders 500 years ago be a pretty big blip on their radar? In a larger sense, what happened when the Europeans invaded North America, Central America, South America and other places was perhaps the largest crime in all of history and it took over 400 years to unfold. If the ancient Maya had some great psychic power that we can benefit

6

from today, why didn't they use it back then to help themselves? This question has tremendous ramifications. For those people who favor the idea that the Maya are predicting a shift into a new golden age, this question offers a difficult challenge. And it is important to also note that even when the Spanish conquest was just around the corner, the Maya still did not seem to benefit from any psychic foresight at all.

In any event, it is very rare for psychics to look at such distant time periods. Remote viewers have tried to look far ahead but they simply don't receive any precise dates. Almost all psychic predictions are basically concerned with the here and now and interest drops off rapidly as the timeline moves away from the present. How many people do you know who are concerned with what is going to happen thousands of years from now?

As I mentioned before, I don't think that the ancient Maya themselves created the calendar to predict world peace at this time. *But this does not mean that we can't create a beautiful world right now!* Your desire for world peace arises from your strong feelings of compassion and love and yet we should be honest and acknowledge that we do not have the power to make someone else happy; we do not have the power to make someone else be at peace. That is up to them and to them alone.

Yet perhaps your attitude and action will inspire others to create deep peace and lasting joy within themselves. In this way, we all can work towards a more peaceful world. Your inspiration and offer of a helping hand to build a better world together can reach out in ways whose limits are unknown. So never give up hope and always do your best to transform both yourself and the world around you into yet an even more beautiful reflection of the divine love that comes to us all from the Source of everything.

It is my opinion that even though some short-term psychic predictions can be genuine and accurate, most of them are not very reliable. I think that this is because there is too much free will in play. This leads me to believe that any prediction that has a timeline of over two thousand years would have little or no merit. It is just not our nature to know what the future will bring and this is especially true when it comes to vast spans of time. I have heard it said that the world was made round so that we would not see too far down the road. While I do believe that psychics and shamans can get a feel for what might be coming our way, these predictions are usually focused on short timelines and rarely guaranteed.

Yet when it comes to astronomy, we can look well into the future and determine exactly what will happen on a particular day! Yes, within the boundaries of the precision of our knowledge, astronomy is guaranteed 100 percent! And you will soon see that there is a simple, obvious and natural astronomical reason why the Maya were drawn to a single day that was over 2,000 years in their future.

So let's now take a closer look at the amazing astronomy of 2012.

Chapter 2
The Basic Astronomy of 2012:
The Triple Rebirth of the Sun and the
Special Configuration of the Sacred Tree

As I mentioned before, don't worry about this being too hard to understand; this is going to be more like a fun art class than a baffling astronomy class! In the overview, you saw a picture of what will be in the sky over the Maya *and over all of us* on that special day. This is the most important picture and in just a few minutes, you will easily understand why this image was used as the foundation of the Maya's transformation and rebirth metaphor. You can do this!

But even still, some of you might want to skip some of this material and that is perfectly fine. I would encourage you to try your best and simply skip anything that seems unclear. I hope you all can at least read the part about the triple rebirth of the sun since this will give you an excellent understanding of the central metaphor of 2012.

So let's get started by talking about what the Maya could see in the sky.

Naked-Eye Astronomy

When we get away from the light pollution of our cities, we all can still view the night sky in the same way as the people who lived long ago. While most people today are only somewhat aware of all the spectacular things that you can see without a telescope, almost all ancient cultures had astronomers who were very familiar with a wide variety of beautiful and magnificent nighttime objects. One of the most obvious and splendid things that captured the attention of all ancient people was the white band of the Milky Way.

As long as we have a clear, dark sky, we can see part of the white band of the Milky Way virtually every night of the year. At different times of the year, we see different sections of this band. The part that we see during the summer months is brighter than the part that we see during the winter months. As the year unfolds, we see that this band actually makes a circle that goes all the way around us.

Now, it was not necessary for the ancient Maya to understand that our galaxy is shaped like a disk or understand our location in that disk. Even though our position in our galaxy is the reason that we see it the way we do, the Maya did not need to understand all that. Please note that I am not saying that they did not know this; I am only saying that they did not *need* to know this in order to make the calendar. Yet they clearly knew that it made a ring that goes all the way around us because that was what they could see with their eyes. The important point is that they often created beautiful metaphorical stories about the wonderful things that they saw with their eyes and so it was with the Milky Way.

While these stories were not meant to be factually true, they were surely intended to teach something about life's deepest mysteries. And one of those mysteries is this: Where did our world come from? Yes of course, the puzzle of creation has been on people's minds since the very beginning. And along with that question is this: Where did *we* come from? *What really happened to us when we were born?*

The ancient Maya were in awe of life. They were in awe of nature and they were infinitely curious about the rhythms of nature, which they observed closely and celebrated throughout the year. It is this wonder and awe that was behind what might be the Maya's most important metaphor and it relates directly to 2012. Let's take a look at it now.

The Sacred Goddess and Her Pregnant Belly of Creation

Night after night, during the summer months, the Maya clearly saw the most spectacular section of the Milky Way. Here they saw a massive ball of light with a dark swath intruding into it. This section of the sky contains the center of the galaxy and that is why it is so bright. There is nothing else in the night sky that looks even remotely like this section of the sky. It is visually quite stunning and very intriguing. The Maya were drawn to it and stories were created about it.

Looking towards the center region of our Milky Way galaxy

No telescope needed

2012essays.com

The ancient Maya saw the band of the Milky Way as the body of the Sacred Goddess and the bulging ball of light as her pregnant belly. The dark rift was the sacred birth canal, the point of creation, the place where the spiritual world transforms itself into the physical world. Here we find the portal between the physical and the spiritual.

This is a very beautiful metaphor! It focuses perfectly on the magic of the human womb, a place where something fantastic seems to appear out of nothing. It takes that miracle that we all have been a part of and expands it to infinity! Metaphorically, everything is said to come into existence from this point of origin. As we shall soon see, this is the section of the sky that is most relevant to 2012.

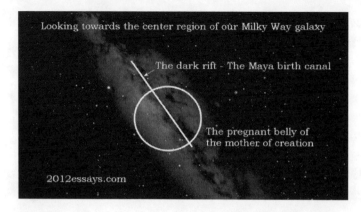

Looking towards the center region of our Milky Way galaxy

The dark rift - The Maya birth canal

The pregnant belly of
the mother of creation

2012essays.com

The Driving Question Behind the Creation of the Long Count Calendar: The Triple Rebirth of the Sun

Almost all ancient people saw the daily sunrise and the winter solstice as two different types of solar rebirths. Did the Maya see yet another type of solar rebirth?

Due to the orbit of the earth around the sun, the sun appears to travel across the dark rift, the sacred birth canal, once a year. The Maya metaphorically saw the one day of the year when the sun was "seen" in front of the middle of the sacred birth canal as a third type of solar rebirth. Let's take a look:

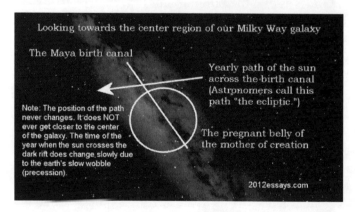

Looking towards the center region of our Milky Way galaxy

The Maya birth canal

Yearly path of the sun across the birth canal (Astronomers call this path "the ecliptic.")

Note: The position of the path never changes. It does NOT ever get closer to the center of the galaxy. The time of the year when the sun crosses the dark rift does change slowly due to the earth's slow wobble (precession).

The pregnant belly of the mother of creation

2012essays.com

Two thousand years ago, the Maya noticed that the sun appeared to travel in front of the dark rift in late November. It did this once a year and each year it did it a little bit closer to the day of the winter solstice. This shift was too small for the Maya to have noticed from one year to the next, but about 70 years later, this difference had accumulated to the point were it was big enough for the Maya to easily detect. Since the occurrence of these two rebirths were slowly getting closer together, the Maya asked the obvious question:

What year in the distant future will have the sun in front of the dark rift on the same day as the winter solstice?

This is the driving question behind the creation of the Maya Long Count calendar! You can already easily see why this event can be seen metaphorically as a triple rebirth of the sun. I call these three rebirths:

1) the daily rebirth
2) the solstice rebirth and
3) the galactic rebirth.

A triple rebirth of the sun happens when all three rebirths happen on the same day.

So another way to state the driving question is to ask, "What year in the distant future will have a triple rebirth of the sun?" And the Maya were indeed able to answer this interesting question. Their careful observations and calculations showed them that a triple rebirth of the sun would happen repeatedly in the years around 2012.

This is the obvious and natural way in which the Maya were attracted to a point in time over two thousand years in their future. This is the main reason that the Maya created their Long Count calendar. Later you will easily learn about the stunning precision that was required to make this calendar. And yes, I do mean stunning!

Before we go on, I would like to clarify a bit about my choice of words. I sometimes use phrases such as "looking towards the center region of the Milky Way." Perhaps the Maya would instead say, "looking towards the belly of the Sacred Goddess" since they may not have had a concept of our galaxy or any idea of where its center region was located. Yet no matter what, they were talking about that section of the sky. Similarly, when I say "the galactic birth canal" you should really hear "the birth canal of the Sacred Goddess." Obviously, the phrases used by the Maya would emphasize their metaphor; they would not use our current astronomical phrases yet I will do so since people today will understand them better.

Now, for those of you who want a better understanding of why the sun appears to travel across the dark rift, let's take a closer look at the way it works.

Earth's Orbit Creates the Galactic Rebirth of the Sun

Throughout the year, due to the orbit of the earth around the sun, from the viewpoint of earth, the sun appears to be continuously moving slowly against the background stars and it is this apparent motion that creates the visual illusion of the sun crossing the dark rift since the dark rift is part of the fixed background of the stars.

Here is a diagram that shows what's involved:

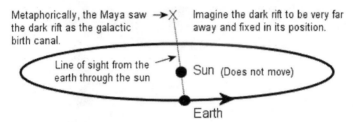

Earth's Orbit Causes Galactic Rebirth of the Sun

Imagine the sun in the middle of a big tabletop with
the earth always on this tabletop as it orbits the sun.

Metaphorically, the Maya saw →✕   Imagine the dark rift to be very far
the dark rift as the galactic   away and fixed in its position.
birth canal.

Line of sight from the →
earth through the sun   ● Sun (Does not move)

Earth

When viewed from above, the earth's orbit is counterclockwise.

Once a year, the earth obits past a position where you can draw a line from
the earth through the sun and into the middle of the dark rift. This creates
the visual image of the sun being in the middle of the "galactic birth canal."

And here is another diagram that helps explain it:

The Earth's Orbit Causes the Sun's
Apparent Motion Across the Dark Rift

Day 1, Day 2 and Day 3 are the   ✕ ←——— Dark rift. Imagine it to be very far
three days under consideration.   away and fixed in its position.

Day 1 - The sun is seen to the   Lines of sight from the earth
right of the dark rift.   ← to the dark rift

Day 2 - The sun is seen in the   Sun (Does not move)
middle of the dark rift.

Day 3 - The sun is seen to the   ● ● ● → Earth's orbital motion
left of the dark rift.   Day Day Day
                                        1   2   3

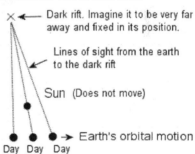

In this view, the earth moves to the right and this creates the visual
illusion that the sun is moving to the left across the dark rift. But the
sun is not moving; it is just a visual illusion. Only the earth is moving.

Here are the screenshots from my astronomy program that show the sun crossing the dark rift in 2012. Each day, the sun appears to move a bit more to the left against the background stars.

Here is the first day:

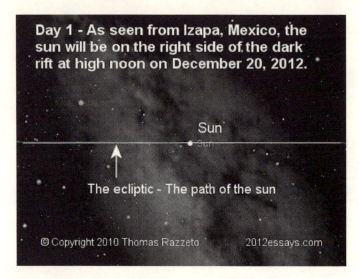

Here is the second day:

And finally, here is the third day:

See how precise this is? Just one day makes a difference in the perfect timing of the galactic rebirth of the sun!

The following simple exercise will help you correctly understand the "motion" of the sun across the dark rift. Hold your left arm straight out with your thumb straight up. Your thumb will be like the dark rift. Imagine it to be very far away. Now, close one eye and hold your right index finger straight up about halfway between your thumb and your open eye. Make it so that your finger appears just to the right of your thumb. The tip of your finger will be like the sun. Now, without moving either your thumb or your finger, move your eye from your left to your right. Your eye is like the earth orbiting the sun. You will see your finger appear to move from the right side of your thumb to the left side of your thumb.

This is just like the sun crossing the dark rift. At the middle of the crossing, your eye is perfectly lined up with your finger and your thumb. While this alignment is real, it is an illusion that your finger moves across your thumb; only your eye is moving, not your finger. So it is with the galactic rebirth of the sun; only

17

the earth is moving, not the sun. Since this point is so important and so often misunderstood, I want to be perfectly clear: the sun does not actually cross the dark rift, it only appears to do so from our viewpoint on earth. But it is this viewpoint that gives us our experience and it is our experience that is so important to us! *This is true in the deepest way about all of our experiences!*

While the rebirth of the sun was a very important metaphor for the ancient Maya, they also had another metaphor for the sun crossing this region of the Milky Way. It was seen as Father Sun mating with the Sacred Mother. In 2012, this special lovemaking will result in the birth of what the Maya called "a new world age," which is represented by a new cycle of the Long Count calendar. Yes indeed, lovemaking and creation are at the core of 2012, not death and destruction.

Before we go on, please note that while the sun will be in front of the dark rift, it will "look" like it will be "in" it so that is how I often state it. But don't get the wrong idea! The sun will not be engulfed in darkness! These stars will simply provide the very distant background for the sun. We are in no danger of any kind because of this annual event!

The Sacred Triple Rebirth of the Sun

Now let's ponder more deeply the astronomy of 2012 and get a better feel for how it provides an excellent foundation for the Maya's metaphor of the sacred triple rebirth of the sun.

Everyday when the sun sets, it goes below the horizon, seemingly under the ground of the earth, and we are left to endure a dark, cold night. Metaphorically, it can be said that the sun leaves our world and travels into the underworld, where it is said to be dead for the duration of the night. Yet at dawn, it rises above the ground and is reborn into our world bringing forth the light and heat we all need to stay alive. Obviously, if the sun were to no longer rise, all of life would perish and because of

18

this, the sun has been used for thousands of years by cultures all around the world as a symbol of the Creator's sustaining love.

Since the sunrise is so frequent, it is often taken for granted, but all of us of course know that this rebirth is truly vital. In addition to that, sunrises are often quite beautiful. In fact, this astronomical shift can be one of nature's most dramatic display of beauty with its blaze of red and pink, yellow and gold. This is the shift that naturally awakens us from our slumber, which, by the way, can also be seen as our own daily rebirth since we, too, appear to be dead while we sleep through the night.

Next, let's consider the winter solstice. It can be seen as the rebirth of the sun in the time frame of the year since the length of the day will now start to grow longer. If the days were to continue to grow shorter, the cold winter would only tighten its grip and we would all perish. So this rebirth is also vital.

Our comfortable homes insulate us from the long, harsh winter nights but when you think of both the daily and yearly rebirths from the perspective of cultures with significantly less physical comfort, you can appreciate that these rebirths of the sun would be experienced in a very tactile way. You can certainly understand the strong motivation to celebrate these turning points in the rhythm of nature.

As we have just learned, the third rebirth occurs when the sun moves in front of the middle of the galactic birth canal, the dark rift, and again this can be referred to as the galactic rebirth of the sun. It also happens once a year but it does not produce any tactile changes as far as our seasons of the year or anything else.

By the way, the solstice rebirth happens once every tropical year and the galactic rebirth happens once every sidereal year. A sidereal year is about 20 minutes longer than a tropical year and this difference is caused by the slow, steady wobble of the earth's axis. This is what causes these two rebirths to come together in the years around 2012.

Now, let's add a little more to our desired triple rebirth scenario. The Maya would most certainly want to celebrate the sun's presence in the middle of the galactic birth canal at a time when this will actually be happening in the sky above them – in other words, during the day. Since the sun is thought to be dead at night, if this astronomical event happened at night, it would not be very useful for their metaphor or very interesting for the purpose of their festivities; they would want the sun to be alive during the rebirth celebration!

So the driving question now becomes: when will the sun be in the middle of the dark rift on the day of the winter solstice while in the sky over the Maya? If there is more than one year that fits these qualifications, what year will also include an interesting planetary configuration? This leads us to the sacred tree.

The Sacred Tree

The sacred tree is in the Maya folklore and other cultures around the world also concern themselves with it. It has several names such as "the sacred cross," "the tree of life," "the sacred tree of life," and "the world tree." The Maya's sacred tree has been shown to be an astronomical reference to the place in the sky where the dark rift is crossed by the path of the sun, as seen from earth. This is true for other cultures as well. Note that this definition is concerned with a place in the sky and as such, it is part of the fixed background of the stars. The definition is not concerned with what may or may not be present in that place, such as the sun or the planets. That is what I call the configuration of the sacred tree.

(Do not confuse the sacred tree with "the southern cross," which is a small constellation of stars mostly seen from the southern hemisphere. I also hear people refer to "the cosmic cross," which is also known as "the grand cross," but again, this is not what I am talking about.)

It just so happens that on December 21, 2012 at high noon, the sacred tree will be beautifully oriented in the sky over the Maya. Since the brightness of the sun will obscure the stars of the Milky Way, we need to push a magic button to let us see the Milky Way, the dark rift and the planets near the sun at this time. When I used my astronomy software program to see this, I was quite stunned. Here (again) is what I saw:

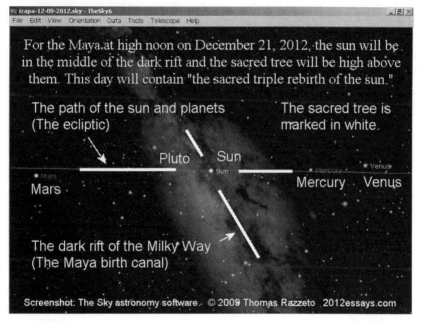

The sacred tree will be above the Maya with the sun evenly between Mars and Venus while in the middle of the dark rift, the Maya birth canal.

At high noon, the sacred tree will contain a large planetary configuration centered around the sun, which will be in front of the middle of the dark rift. This is not some tiny thing; it will take up a large part of the sky. To get an idea of how big this will be, please do the following. If you do this on a day near the winter solstice at high noon, you will get a decent idea of the size and location of the sacred tree on this day. Okay, so hold out both of your hands at arms-length and spread your fingers and thumbs out as far as you can. Place the tips of your thumbs together where the sun is. The tip of your left little finger is where Mars will be and the tip of your right little finger is where Venus will be. The pregnant belly of the Sacred Goddess is about the size of your fist. See how big this is? Fantastic!

If we draw a horizontal line through the sun, we have one of the crossbars of the tree of life. On the left side of this crossbar, we will have the planet Mars. To the right of Mars will be Pluto and then the sun. To the right of the sun will be Mercury and then at the far right of this crossbar, we will have Venus. Since Pluto is never visible to the naked eye, the Maya were completely unaware of it; I only labeled it since it is actually there. But the ancient Maya were certainly very familiar with the other three planets and precisely tracked their position, as seen from earth.

Now, I call this horizontal crossbar the crossbar of light since all these objects either reflect or give off light. The sun will be virtually in the middle of Mars and Venus so this crossbar will be very balanced around the sun. The dark rift makes what I call the crossbar of darkness. At midday, this crossbar is not exactly vertical but it is 30 degrees shy of vertical.

In this way, we see that the tree of life is made up of one crossbar of darkness and one crossbar of light and thus it represents the yin-yang quality of duality exhibited throughout all of creation. And right on the crossing point of these two crossbars on this special day, we have the sun, which brings us the light and warmth of the day when it rises and the darkness and coolness of the night when it sets. Because of this, the sun can be seen as the symbol for the powerful creator of the duality that we witness in our world. It even seems to bring us life itself and invites us to inquire about the most mysterious duality we all face: our very own life and death.

When we take all of this into consideration, we see that there is absolutely no doubt that December 21, 2012 offers an excellent solution to the driving question behind the creation of the Long Count calendar!

As I looked at the screenshot above, I realized that throughout this day, Venus will lead a parade across the sky starting before dawn and that Mars will bring the parade to a close about one hour after sunset. Of course the sun will be the guest of honor since it is by far the brightest object in the sky and this is the day of its magnificent triple rebirth. And Venus is the perfect leader of this parade since it is the brightest of all the planets. The Maya would certainly have been aware of the spectacular nature of this parade!

(Please note that basically this parade will be seen where you live with all the times shifted to match your location. Of course, if you live in the southern hemisphere, you will have a summer solstice, not a winter solstice. As such, it might seem that the triple rebirth metaphor won't apply to you but I don't think that this is really a problem at all. In my opinion, as you will see later, the Maya's transformation and rebirth metaphor is about the fundamental nature of reality. As such, this metaphor is not only timeless, but it also applies to everyone all around the world.)

So let's check out the details of this fabulous parade.

We will be using the time for Izapa, Mexico, which is in Central Time, since Izapa is the birthplace of the Long Count calendar.

What Will Actually Happen on December 21, 2012: Venus Will Lead a Spectacular Parade Across the Sky

As seen from Izapa, Mexico, Venus will start things off by rising above the horizon at 4:46 AM. It will be extremely easy to see in the black predawn sky and it will lead the sacred tree on its journey across the sky throughout this special day. At 5:12 AM, the moment of the winter solstice will arrive and the sun will be reborn with the days now becoming longer. Next, Mercury will rise at 5:23 AM and it will be visible even with dawn's increasing light. It will be the second object on the sacred tree. A little more than an hour later, the sun will enter our world with a blaze of color at 6:29 AM to become the most important object on the sacred tree, the guest of highest honor! Shortly after this, the sky will become so bright that both Mercury and Venus will be obscured from view but nonetheless, they will continue their journey across the sky with Venus leading the way. Pluto, which is never visible to the naked eye, will rise at 7:03 AM and finally Mars will rise at 8:23 AM, making it the last object on the sacred tree and this is why it will be the one to bring the parade to a close after sunset. Notice that the sun will rise 103 minutes after Venus and that Mars will rise 115 minutes after the sun. This means that the sun will be very close to the middle of these two planets.

Let's take a look at some screenshots.

Here it is just after sunrise:

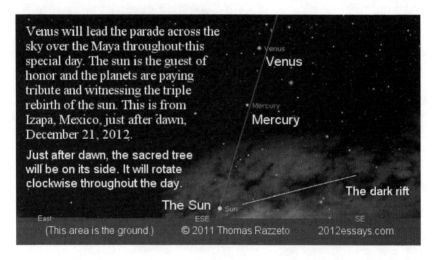

Venus will lead the parade across the sky over the Maya throughout this special day. The sun is the guest of honor and the planets are paying tribute and witnessing the triple rebirth of the sun. This is from Izapa, Mexico, just after dawn, December 21, 2012.

Just after dawn, the sacred tree will be on its side. It will rotate clockwise throughout the day.

The dark rift

The Sun

Venus

Mercury

East ESE SE

(This area is the ground.) © 2011 Thomas Razzeto 2012essays.com

And here it is shortly after Mars rises:

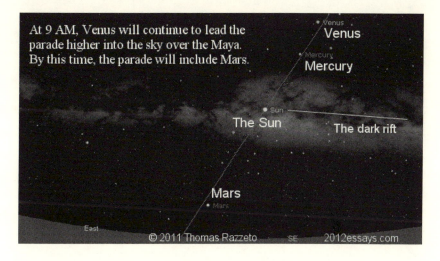

Here it is a little later in the morning:

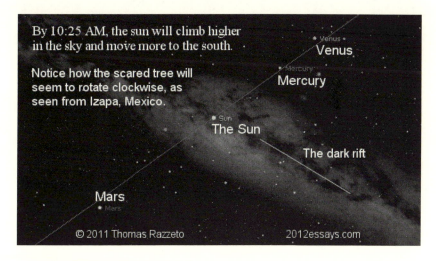

And here again is the sacred tree at high noon:

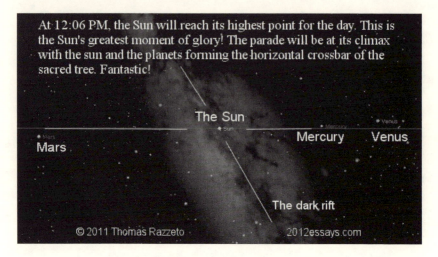

At 12:06 PM, the Sun will reach its highest point for the day. This is the Sun's greatest moment of glory! The parade will be at its climax with the sun and the planets forming the horizontal crossbar of the sacred tree. Fantastic!

The Sun

Mars

Mercury

Venus

The dark rift

© 2011 Thomas Razzeto

2012essays.com

Here it is in the early afternoon:

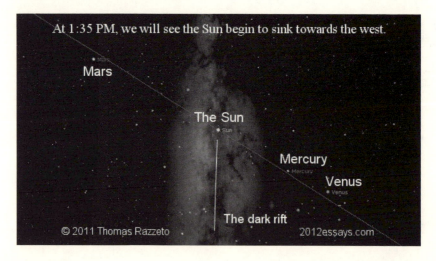

At 1:35 PM, we will see the Sun begin to sink towards the west.

Mars

The Sun

Mercury

Venus

The dark rift

© 2011 Thomas Razzeto

2012essays.com

And here it is about two hours later:

At 3:30 PM, we will have Venus leading the parade towards the western horizon. The parade will soon start to end.

Notice how the sacred tree will continue to rotate clockwise.

Mars

The Sun

The dark rift

Mercury

Venus

© 2011 Thomas Razzeto   2012essays.com   West

And finally, here it is about a half-hour before sunset. The parade is coming to a close:

By 5:15 PM, Venus and Mercury will have set and the Sun will soon follow. By about 7:40 PM, Mars will set and the parade will be over.

Mars

The Sun

SSW   SW   WSW   West
(This area is the ground.)   © 2011 Thomas Razzeto   2012essays.com

As we can see in the previous screenshots, as the morning progresses, the sacred tree will rise higher in the sky and move from the eastern horizon to its highest point in the southern sky. As you may have already figure out, the sacred tree will be in its most ideal orientation during the middle three or four hours of the day since it will be high in the sky and reasonably upright.

For the rest of the afternoon, the sacred tree will sink slowly towards the western horizon. As it does so, the sun will continue to shift slowly against the background stars. It looks to my eye like the sun will be virtually exactly in the middle of the Maya birth canal at about 4:30 PM. Let's zoom in for a closer look.

December 21, 2012, 4:30 PM, Izapa, Mexico - The sun will be in the middle of the dark rift of the Milky Way.

Sun

The ecliptic - The path of the sun tilts downward late in the afternoon.

Screenshot from The Sky, a popular astronomy program. Copyright 2009 Thomas Razzeto   2012essays.com

This is it! The position of the sun in the middle of the dark rift, the Maya birth canal, on the same day as the winter solstice plus the unique configuration of the sacred tree are the reasons the Maya picked this exact day!

But let's continue since there is still a little bit more. At 5:44 PM, the sun will set and Mother Nature herself may provide another spectacular scene of glorious colors. Finally, at 7:38 PM, Mars will be visible against the night sky as it sinks below the horizon, marking the end of the sacred tree's journey across the sky on this fantastic day. I can only imagine that the Maya celebration will continue throughout the night! Notice how the Maya will be able to see Venus and Mercury before dawn and Mar after sunset. So even though they will not be able to see these planets on the sacred tree during the daytime, they will be able to see them at some point during this calendar day.

What the Maya can actually see with their eyes is crucial in understanding 2012.

As we learned earlier, the position of the sun in the middle of the dark rift on the same day as the winter solstice is the reason the Maya looked towards the years around 2012. But it is the special configuration of the sacred tree on the day of the winter solstice of 2012 that completes the story about why the Maya picked this exact day. Some 2012 researchers talk about a "zone" of years and this is because they ignore the special configuration of the sacred tree. It is this additional consideration that "snaps" the date of interest from a zone of years to this precise day!

I like to think of it like this. If you are in a huge wilderness area and you find a road and you follow it and it suddenly comes to an end, you may wonder why it ended at that particular spot. If you look up and see a beautiful lake, you can speculate with reasonable certainty that the lake is the reason the road ends where it does. But if instead there is no lake and everything around you looks the same as the general area, it might remain a mystery as to why the road ends where it does. Yet the Maya calendar is not like that. On the exact day that the calendar restarts, this spectacular astronomy is found in the sky over the Maya. *This is the beautiful lake! After several years of research, this for me is clearly why the Maya restart their calendar on this exact day. In my opinion, this is beyond a reasonable doubt.*

By pointing to the sacred triple rebirth of the sun, the Maya are pointing to cycles within cycles. Yes, these natural cycles are the rhythms of nature that bring us life itself and they will repeat forever. But by pointing to this beautiful and unique configuration of the sacred tree on this special day, the Maya are reminding us that each and every moment is unique and precious even though they arise within cycles. If you only consider the presence of the sun in the middle of the birth canal on the day of the winter solstice, you will see that this pattern of the triple rebirth of the sun will occur again 26,000 years from now. But when you consider the position of the sun and the planets on the sacred tree, you have an event that will never repeat; it is totally unique!

If you only focus on the repetitive patterns of nature, you might think that you are stuck like a cog on a wheel in a big machine that plays itself over and over again. But when you realize that within these vital cycles, each moment is unique, you will realize that you are alive and dynamically participating in a unique way in the mystery of life at every moment. Yes, every moment is sacred and the special configuration of the sacred tree on this exceptional day contains a beautiful built-in reminder to that effect.

Before we move on, let's just have a little fun for a minute and listen to two ancient Maya talking about 2012.

Skit About Two Ancient Maya Talking About 2012

"Hey, have you heard? There is gonna be a really fantastic thing up in the sky!"

"Fabulous! I love really cool things like that. When's it gonna be?"

"Oh, it won't be for a while yet. The guys are telling me it'll be in about ... oh ... two thousand years. But they know the exact day!"

"Wow! That's stunning! But ... that's a really long time from now."

"Well, yeah, ... but it's super important and I don't want you to miss it. So put it on your calendar!"

The point of the skit is again the very unusual focus on a point in time over two thousand years in the future. Why would anyone care? I think that they would only do this because of the *certainty* of the actual astronomy of that day and because of the *timeless* nature of their metaphorical message.

It is the timeless nature of the message that makes it so profoundly relevant to the Maya so long ago.

You'll learn more about the message of 2012 in chapter four and most of you should just skip ahead to that chapter right now. But for those of you who really want to learn all about the astronomy of 2012, please continue with chapter three.

Chapter 3
More Astronomy

Again, this chapter is only intended for people who really want to dig deeply into the astronomy of 2012. Yet perhaps you will give it a try since it really isn't that hard to understand if you go slowly and take it one step at a time. Maybe you'll surprise yourself!

The Configuration of the Sacred Tree During Other Years

Do you remember that I said that there are many years around 2012 that have a triple rebirth of the sun? Just out of curiosity, let's take a quick look at some of them. Here we will concern ourselves with the planets near the sun on the day of the winter solstice. As you know, this is what I call the configuration of the sacred tree. Each year, the configuration will be different.

Here is the configuration on the winter solstice of 2008:

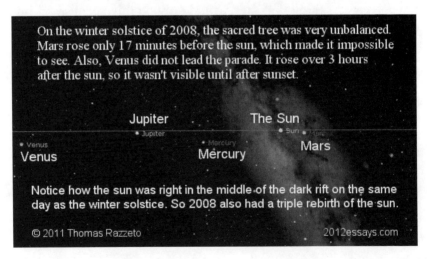

On the winter solstice of 2008, the sacred tree was very unbalanced. Mars rose only 17 minutes before the sun, which made it impossible to see. Also, Venus did not lead the parade. It rose over 3 hours after the sun, so it wasn't visible until after sunset.

Jupiter
The Sun
Venus
Mercury
Mars

Notice how the sun was right in the middle of the dark rift on the same day as the winter solstice. So 2008 also had a triple rebirth of the sun.

© 2011 Thomas Razzeto   2012essays.com

And here it is for 2009:

On the winter solstice of 2009, the sacred tree only had Venus and Mercury. Venus rose about 20 minutes before the Sun and was very difficult, if not impossible, to see with the naked eye.

The Sun

Mercury

Venus

Notice how the sun was right in the middle of the dark rift on the same day as the winter solstice. So 2009 also had a triple rebirth of the sun.

© 2011 Thomas Razzeto

2012essays.com

Here is 2010:

On the winter solstice of 2010, the sacred tree only had Mercury and Mars. Mercury rose about 15 minutes before the Sun and was impossible to see with the naked eye.

The Sun

Mercury

Mars

Notice how the sun was again in the dark rift on the same day as the winter solstice.

© 2011 Thomas Razzeto

2012essays.com

33

Here's 2011:

On the winter solstice of 2011, the sacred tree only had Mercury and, of course, the Sun. Mercury rose about 90 minutes before the Sun and was visible without a telescope in the predawn sky.

The Sun
· Sun

· Mercury
Mercury

Notice how the sun was again in the dark rift on the same day as the winter solstice.

© 2011 Thomas Razzeto

2012essays.com

We have already seen 2012 so let's look at 2013:

On the winter solstice of 2013, the sacred tree will only have Mercury and Venus. Mercury will rise 17 minutes before the Sun, which will make it virtually impossible to see. Venus will rise about 2 hours later, which means that is will only be visible after sunset.

The Sun
· Sun

· Venus
Venus

· Mercury
Mercury

Notice how the sun will again be in the dark rift on the same day as the winter solstice.

© 2011 Thomas Razzeto

2012essays.com

And here is 2014:

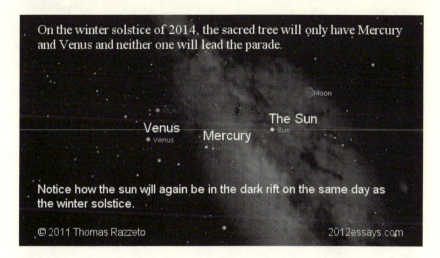

On the winter solstice of 2014, the sacred tree will only have Mercury and Venus and neither one will lead the parade.

Moon

Venus   The Sun
Mercury
● Venus   ● Sun
● Mercury

Notice how the sun will again be in the dark rift on the same day as the winter solstice.

© 2011 Thomas Razzeto   2012essays.com

And finally, here is 2015:

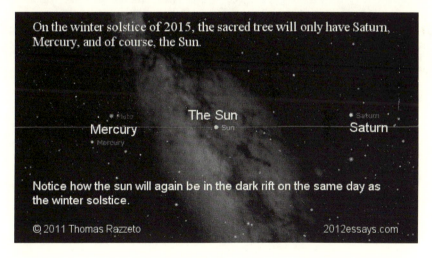

On the winter solstice of 2015, the sacred tree will only have Saturn, Mercury, and of course, the Sun.

The Sun
Mercury   Saturn
● Sun
● Mercury

Notice how the sun will again be in the dark rift on the same day as the winter solstice.

© 2011 Thomas Razzeto   2012essays.com

I could go on and on with other examples but the point is that the Maya picked 2012 because of the visibility, balance and beauty of the sacred tree. I know that it doesn't sound very scientific, but I think that one of the reasons that the Maya picked 2012 is because the sacred tree will be so beautiful in that year.

35

All the examples have the sun in the Maya birth canal on the same day as the winter solstice. Of course we know that this is what makes the triple rebirth of the sun. If the Maya were only concerned with the triple rebirth of the sun, they could have picked any of these years along with many others. But they picked 2012. In my opinion, this is like picking a rose from the garden. Some roses are just nicer than the others but they are all still roses. The Maya could have picked any of these years to convey the same message of transformation and rebirth. Was it just luck that the Maya picked a configuration that will be so beautiful? Given the amount of precision demonstrated by whoever created this calendar, it seems to me that they were also aware that in 2012, the sacred tree would have this beautiful configuration.

By the way, there is another astronomical event that the Maya knew about that may have been an additional reason why they picked 2012. It is something called a Venus transit, which happens when the orbit of Venus takes it directly between the earth and the sun. This means that a small black dot made by Venus can be seen traveling across the bright disk of the sun. This can sometimes be safely seen by the naked eye without optical filters at dawn and dusk, if you have the right cloud conditions.

Because the orbit of Venus around the sun is not exactly in the same plane as the orbit of earth around the sun, Venus will often travel just above or just below the disk of the sun, as seen from earth, and therefore it will not be seen as a dot moving across the disk of the sun. But sometimes it is seen as such! These events have an unusual rhythm. Here are four occurrences: One happened in 1882 and the next one happened in 2004. The next one will occur on June 5, 2012 and the one after that will not be until 2117. The Maya had a calendar that tracked this so they knew one would occur in 2012.

Four other astronomical events that might have influenced the Maya's choice of 2012 are two solar eclipses and two lunar eclipses. We know from their ancient writings that the Maya predicted eclipses of both the sun and the moon very accurately but since there are on average two lunar eclipses per year and since the solar eclipses will not be visible where the Maya lived, I do not think that these things played a big factor in the selection of 2012.

The Wobble of the Earth

I want to take a minute to cover some additional evidence that the Long Count calendar is based on astronomy. As I mentioned earlier, the earth is slowly wobbling on its axis like a top that is not standing straight up. This is called "precession" and again, it takes about 26,000 years for the earth to complete one wobble. This period of time has several names and one of them is "the great year."

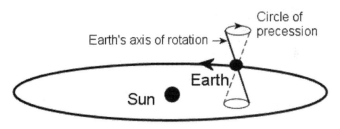

Earth's Precession - It Wobbles Like a Top

The earth completes one wobble every 25,800 years.
The axis of rotation remains tilted by 23.5 degrees.

Circle of precession

Earth's axis of rotation

Earth

Sun

As the earth orbits the sun, it also slowly wobbles. When viewed from above, the earth's orbit is counterclockwise but the direction of precession is clockwise.

© Copyright 2010 Thomas Razzeto   2012essays.com   Not to scale

The Maya Long Count calendar is 5,125 years long and five cycles of the Long Count calendar add up to the length of the great year to within one percent. Since the Maya frequently talk about all five cycles and some of their artwork reflects this awareness, I don't think that this is just a coincidence.

A five-pointed Maya star. Five cycles of the
Maya calendar add up to the great year.
The cycle of Venus also has five parts.

This five-pointed star also refers to a cycle of Venus. I won't go into this too much now, but it involves the ratio of the time it takes Venus to go around the sun as compared to the time it takes the earth to go around the sun. This ratio is reflected in a cycle of Venus called the synodic period and over the course of eight earth years, Venus, as seen from earth, goes through five of these synodic periods. Then the whole thing repeats again. This can be seen with the naked eye and the rhythm can be measured by counting days. This is why Venus is sometimes seen before sunrise as the morning "star" and at other times after sunset as the evening "star." The Maya were very aware of the cycles of Venus and they had a calendar to track it with great precision. This five-pointed star is a good example of one symbol that points to two things and it gives us a clue that perhaps the Maya divided the great year into five parts in order to reflect the natural rhythms of Venus rather than dividing it into four parts to reflect the seasons of the year, which is what I might have expected.

It was the precise correlation between the length of the great year and the length of the calendar plus the fact that the Maya hit the exact day of the winter solstice that initially convinced me that the actual astronomy of 2012 was worth a thorough investigation. The more I dug into it, the more amazing it all became! And if the calendar is not intentionally related to precession, what on earth could it be for? Surely the Maya did not use it to grow their corn! Why would they need to track such a long cycle?

To correctly pinpoint a winter solstice from 2,000 years away shows a precise understanding of the length of the year. And again, the length of the Long Count calendar clearly demonstrates knowledge of the great year. So it seems to me that the Maya are saying,

"Yes, we understand the year and yes, we understand the great year. We have extremely precise values for both."

As a side note, consider that in about 130 BC, the same time frame as the creation of the Maya calendar, the Greek astronomer Hipparchus estimated precession to be 36,000 years or less. He was off by 10,000 years! Even so, Hipparchus is very famous for his work on precession and he wrote two books on the subject. He is considered by many people to be the greatest astronomer of antiquity. I can only add that it's a good thing he was not in charge of creating the Maya calendar!

A Bigger Idea: The Rebirth of the Great Year

I find the triple rebirth of the sun to be fascinating. But perhaps the Maya are actually using it to point to something even bigger: the rebirth of the great year. Without going into the details here, it turns out that we can metaphorically state that the great year also has four seasons, just like our regular year. In this way, we find that the great year will soon be reborn during its own "winter solstice" and that December 21, 2012 is a reasonable choice as the time when this will happen.

Please note that I am offering this idea as a metaphor, just for fun. Our astronomers know that there are no physical effects on the earth due to the angle of earth's axis with respect to the background stars. And I don't believe that there are any spiritual effects either, since there is no historical evidence to support that human consciousness "bobs" up and down with this 26,000-year cycle. And yet, just for fun, we might say that on the day when the calendar itself is reborn, we have the sacred triple rebirth of the sun and the rebirth of the great year! Interesting!

From the viewpoint of the Maya, December 21, 2012 is more than just another winter solstice; it is the most important winter solstice of all the winter solstices throughout the entire great year. This might make this day the single most important day in the entire great year! There are over 9.4 million days in the great year and the Maya have picked this day as the day when the great year is reborn!

The Maya were aware of the effects of precession. The calendar's restart date undeniably contains all this amazing astronomy that is keyed to precession and the position of the planets. The poetic beauty is breathtaking; the actual astronomy is profoundly precise! How can all this be just a coincidence? I find it to be completely mind-blowing!

To my knowledge, I am the only researcher who describes the special event of 2012 exactly like I do. Some researchers focus on 1998 and the years surrounding it. But I don't think that that is what the Maya were trying to pinpoint with their calendar, though missing it by 14 years. I think that they successfully hit exactly what they were aiming at: the sacred triple rebirth of the sun unfolding in the sky over the Maya!

The Maya Had An Extremely Precise Value for the Length of the Year

Before we go into the question of how the Maya measured the length of the year, I want to show how they measured the length of the cycle of the moon. While the cycle of the moon does not come into play for 2012, I would like to offer it as an example of the level of precision that the Maya achieved by simply counting days. Let's check it out.

We all know that the amount of time that it takes the moon to go from a new moon to the next new moon is about 30 days. Our modern astronomers have measured this value to be 29.5306 days. In Copan, the Maya recorded that they observed that the moon went through 149 of these cycles in precisely 4,400 days. When you divide 4,400 by 149, you get 29.5302 days. Excellent! This is about 99.999 percent correct! And notice that they had the dedication to spend about 12 years studying a cycle that was only about 30 days long! In this simple example, we see that the Maya were able to do some impressive work without telescopes, clocks or computers. It is very important to note that measuring time by just counting days worked great!

The Maya were only concerned with what they could see with their eyes and they measured time by counting days.

I cannot stress this enough! And in this regard, we must not overlook the fact that the Long Count calendar itself inherently counts days! Okay, with this example in mind, let's take a look at how the Maya measured the length of the year.

Our modern astronomers have measured the length of the year as 365.2422 days. This converts to 365 days, 5 hours and 48 minutes. Now let's imagine that the Maya knew the length of the year with an error of only 1 minute. Suppose they thought the year was 365 days, 5 hours and 49 minutes long.

Let's say they wanted to project 10 years into the future and exactly pinpoint the day that will include the moment of the winter solstice. (Yes, the winter solstice is a moment in time, not a whole day as some people incorrectly think.) When the Maya project out 10 years, their 1-minute per year error would accumulate into 10 minutes. But this would probably not be a problem since they only want to know the date that contains the moment of the winter solstice. Although they would be off by 10 minutes, they would probably still be on the correct date. Notice, however, that if the solstice turns out to be 5 minutes before midnight, they would incorrectly calculate it to be 5 minutes after midnight and they would therefore pick the wrong date. This effect will always be a minor concern, no matter how far they project, short or long, and no matter how precisely they know the year. But let's not worry about this and just continue.

Now let's consider projecting out 100 years. In this case, the total error would grow to 100 minutes and again, this would probably still be okay since it is an error of less than 2 hours. But since the Long Count calendar projects over 2,000 years into the future, the error would grow to over 2,000 minutes, which is longer than the length of the day, which is 1,440 minutes. This means that they would surely miss their target date. So we see that even an error of only 1 minute per year is too large! Now that you see the general problem, let's bring in more precision.

What we are doing is simply noting that our tolerance is a total error of one day for the entire 2,000-year period that started when the Long Count calendar was first put into use. We do not want to be one day too late or one day too early. We want to hit the exact date that contains the moment of the winter solstice.

To calculate our allowed error per year, all we need to do is divide 1 day by 2,000 years. If we want our answer in minutes, we should divide 1440 minutes by 2,000 years to get 0.7200 minutes per year. If we want our answer in seconds, we multiply this by 60 seconds per minute, which gives us: 43.20 seconds per year. I often round this to simply 45 seconds.

When I first did this calculation, I was so amazed that I could hardly get it out of my mind. For days, I would stop and say to myself, "45 seconds! 2012 points to the triple rebirth of the sun and the Maya needed to know the length of the year to within 45 seconds!"

Now let's continue by putting this into a percentage. First we need to get the total number of days in our 2,000-year period. So we multiply 2,000 by 365.2422, which gives us 730,484.4 days. A one-day difference is what percentage of this value? Or, as I prefer, let's just trim one day off this total and see what percentage we have left. As most of us remember, to calculate a percentage, we multiply by 100 and divide by the total:

(730,483.4 x 100) / 730,484.4 = 99.99986, which I round to 99.9999 percent. Wow! The Maya needed to know the length of the year to a very precise level!

This is like measuring the width of the United States to within 20 feet or the distance from Los Angeles to Tokyo to within 40 feet! In my opinion, this is not merely remarkable; this is absolutely stunning!

Now that you know the amazing precision that was required, you are probably asking yourself, "How in the world did they achieve this?" So let's dig into that question.

How the Maya Measured the Length of the Year

The goal is to measure the length of the year. For our purposes, we will define the year as the amount of time from the moment of one winter solstice to the moment of the next winter solstice. One approach is to focus on a single year. Here are the three steps of that approach:

1) Accurately detect the physical conditions that create the moment of the winter solstice.
2) Accurately record the time when this happens with a clock.
3) Do this for two consecutive winter solstices and subtract the recorded times.

This approach requires that the Maya use a clock that is very accurate over an entire year. Did they have such a clock? And how would they precisely detect the physical conditions when the axis of the earth is exactly pointing away from the sun, as much as possible, since this is the definition of the winter solstice?

But consider the following multi-year approach, which does not require a clock or the ability to precisely detect the conditions of the solstice. It will provide excellent accuracy for the length of the year. Here's how it works.

Many ancient cultures have used several ingenious methods for detecting the day of the winter solstice. Some of these methods focus on the length of the shadow produced by the midday sun. The day of the winter solstice produces the longest shadow. The method that I will discuss here is just a bit different in that it uses a beam of sunlight rather than a shadow. Here's how it works. If we construct a building with a round window in a wall that directly faces the sun at midday, it will create a beam of sunlight that will trace a path on the floor as the sun moves across the sky. This path is different for each day of the year and we will pay the most attention to it in the few hours around high noon. While this method does not give us the precise moment of the winter solstice, it will reveal the day that contains it. This type of building is called a pinhole solar observatory even though the window can be several feet or more in diameter – hardly what I would call a "pinhole," but that is a minor point. By the way, the bigger the building, the easier it is to measure the details of the path of the sunbeam.

Now comes the part that takes many years. We don't use a clock but instead, we count the days from the day of the winter solstice of our starting year to the day of the winter solstice many years later. For a period of 100 years, for example, we would count 36,524 days and this gives us a value of 365.24 days for the length of one year. Pretty good!

If the building is large enough, some of the subtle details about the paths that are traced on the floor can be put to good use. By paying careful attention to the path of the sunbeam for the days around the day of the solstice, we can get a rough idea as to what part of the day contained the moment of the solstice. For example, if the moment of the winter solstice was at high noon, then the path traced out for that day will stand out very well from the paths traced out the day before and the day after. The paths of those adjacent days would be very similar to each other but they would peak below the position of the day of the solstice. And if no single day has such a strong peak and instead two sequential days have similar paths, the solstice was near midnight between the two days and it would be hard to tell which day contained the moment of the winter solstice. It seems to me that with this attention to detail, it might be possible to get within 6 hours of the moment of the solstice, perhaps even much better.

So we have a 6-hour tolerance at the beginning of our 100-year long measuring period and we also have another 6-hour tolerance at the end of our measuring period. So we have a total tolerance of 12 hours for our 100 year period and this can be expressed as plus or minus 6 hours. Since it is spread out over 100 years, we need to divide by 100. If we convert 6 hours into minutes and divide by 100, we get a tolerance of plus or minus 3.6 minutes. Excellent!

But we have seen that the Maya needed a tolerance of about 43 seconds, which can be stated as plus or minus 22 seconds. So we still need a value approximately 10 times more precise. To improve our results, we can increase our observational period and / or increase our accuracy of detecting the moment of the winter solstice or both. What I found in this regard actually surprised me.

The ancient people of Mesoamerica had plenty of time on their hands. In fact, this is the key that unlocks the puzzle. In chapter 6 of John Major Jenkins' book, *The 2012 Story*, we learn of the work of archaeologist Marion Hatch. While John's information about Hatch's work is very brief and a bit sketchy, Hatch tells us that about 3,000 years ago, people in La Venta, which is a town just north of Izapa, learned of precession when they built a sacred building that pointed to the exact point on the horizon where a certain star rose from. (That star was probably Sirius. It probably attracted their attention because it is the brightest star and can often be easily seen rising in the east.) Yet after about 70 years, the people noticed that the star now rose from a slightly different location and they rebuilt the building to correctly point to it again! The earth's very slow wobble had caused this change. And any change like that would inspire their astronomers to watch closely and measure carefully. This would include counting days.

As you know, we are now discussing the length of the year and not precession. But the important point is that we now know that people in Mesoamerica were counting days going back 3,000 years. This means that they had about 1,000 years to observe the length of the year before they created the Long Count calendar.

One thousand years of observational data is the key!

Now we clearly see that the Maya were not trying to directly measure the length of one year to within a certain number of seconds per se, they were simply counting the number of days from one winter solstice to another winter solstice many, many years later. Knowing the exact number of days in one thousand years will allow them to calculate the exact number of days until the winter solstice of 2012. This approach is just precise enough to make the calendar! The error projecting 2,000 years into the future would be less than one day. Fantastic!

It is so difficult to believe that the Maya could have had this level of precision that it is easy to understand why mainstream science denies them this knowledge and attributes the restart date to mere coincidence. But is this justified? Is the restart date landing on the winter solstice a coincidence?

If the Maya were Christian and the restart date was Christmas, would we be justified in stating that it was just a coincidence? Of course not! It would be obvious that it was intentional. And yet – in a way – the winter solstice plays a role similar to Christmas for the Maya. It was a very important day of the year! It is too much for me to think of this as just a coincidence.

Another thing that bothers some people is this: How did the Maya know which stars were behind and near the sun since the sun's brightness obscures these stars? As you know, for my thesis to be valid, the Maya have to be able to do this with a high degree of precision in order to predict the sun's position in the dark rift, which again is part of the background stars. Let's check it out.

How Did the Maya "See" the Stars Near the Sun?

The first thing to realize is that the stars that form the background for the sun repeat year after year. This is due to the orbit of the earth around the sun. This type of year is called a sidereal year and it is, on average, about 20 minutes longer than a tropical year, which is also called a solar year. A tropical year is determined by the seasons of the year, which, as you know, are caused by the angle of the axis of the earth with respect to the position of the sun. A sidereal year is determined by the background stars. Let's take a look.

The Stars Behind the Sun, As "Seen" from Earth

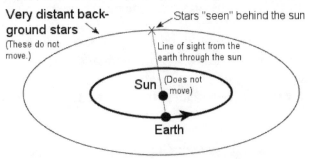

As the earth moves to our right, the sun appears to move to our left against the very distant background stars. But this apparent motion is due to the motion of the earth, not the sun. Every year, the stars behind the sun repeat.

While the sun seems to move smoothly and continuously against the background stars, as seen from earth, this apparent motion is actually caused by the motion of the earth as it orbits around the sun. The earth's orbit around the sun always stays in the same plane. In other words, the earth does not bob up and down as it goes around the sun. Because of this, the background stars that seem to pass behind the sun are exactly the same year after year. By the way, the band of stars that lie behind the apparent path of the sun is called the zodiac. (The band of stars that make up the zodiac should not be confused with the band of stars made by our Milky Way galaxy.)

The zodiac attracted great attention from all ancient naked-eye skywatchers because the sun and the planets can only be observed near this section of the sky. While the planets and the sun do move against the background stars, they are not free to just wander around anywhere at all! They must stay fairly close to the zodiac, which makes a huge circle around the earth. In fact, the sun must stay exactly on something that astronomers call the ecliptic, which also goes around the earth, right through the middle of the zodiac.

So here is what I think happened long ago. Initially, the Maya created a map of all the stars that they could see from their location on earth. Each night they saw a little bit more of the stars in one direction of the zodiac and a little bit less in the other direction. Given good viewing conditions and a fairly low horizon, it only took one year to create a complete map of all the stars that were visible to them.

Incidentally, I have heard that the Maya used reflecting ponds to help them with this. The image of the night sky, which was seen as a reflection in the pond, could have been measured with ropes. In this way, angles, relative size and distance could be reproduced very precisely. The Maya refined their map of the stars over many years to the point where they were very accurate.

Once they had this star map, the Maya could put it to use as follows. Just before dawn, they would take note of the last visible constellation nearest the rising sun before the brightness of the sun made it fade away from view. Later that day, just after sunset, the Maya would take note of the first visible constellation nearest the setting sun. They would then know that the sun was approximately in the middle of these two constellations. But that was just a start. The Maya also used other observations to learn the position of the sun with much greater precision.

It is my opinion that the Maya correctly understood that the earth went around the sun and that the moon went around the earth and that the moon was closer to earth than the sun. This – and many other things – can be determined by logically observing the day, the seasons of the year, the phases of the moon, and eclipses of the sun and the moon. Although the Maya did not have high technology, some of them were certainly very smart!

And when they saw a full moon in a certain position against the background stars, which they could see, they knew that the sun was in the exact opposite position of the zodiac. This is just simple geometry. An eclipse of the full moon is a special case of this geometry. During an eclipse, the shadow of the earth falls on the moon; otherwise the shadow goes just above or just below the moon leaving it free to shine brightly. Since the Maya understood this, they could use these events and their star map to determine where the sun was in relation to the background stars on the days when there was a full moon.

The Full Moon Determines the Stars Behind the Sun

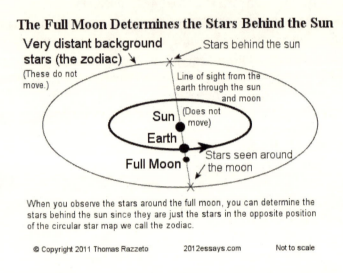

When you observe the stars around the full moon, you can determine the stars behind the sun since they are just the stars in the opposite position of the circular star map we call the zodiac.

While the full moon and lunar eclipses were quite helpful, there was still another event that was even better: a solar eclipse. During a total solar eclipse, the sky gets dark enough for a few bright stars to be seen by the naked eye and this allowed the Maya to directly observe the actual position of the sun against the background stars. This is exactly what they wanted to know! While total solar eclipses do not happen very frequently, they happened often enough that the Maya learned how to predict them and to use them in this way.

Next, let's consider the following. In the last week of February of 2012, I was able to observe Venus, Jupiter, the moon and the sun all at the same time without a telescope. The crescent moon was easy to see throughout the day and Venus became visible about 20 minutes before sunset. I was surprised that I could also see Jupiter just a few minutes before the sun went below the horizon. This allows for a map to be made of the position of these four objects and after the sun sets and the sky gets dark, you can see the position of the remaining three objects against the background stars. This allows you to determine the stars behind the sun very accurately. Even if you can only see a single object such as the moon or Venus while the sun is still up, you can still get a good idea of which stars are near the sun. And of course you can make similar observations near dawn. In this way, the ancient Maya could probably get a good idea of the position of the sun ten to twenty times a month, even if they were only using the visual position of the moon and the sun.

By taking into account an observational period of 1,000 years, we can now understand how the Maya knew the length of the sidereal year and the stars that were near the sun on any given day. They also needed to know the rate of precession and the amount of shift required to bring the galactic rebirth onto the same day as the solstice rebirth. In addition to that, they also needed extremely precise knowledge to accurately predict the positions of the planets on the sacred tree. All of this leads some people to conclude that extraterrestrials created the Long Count calendar. But again, it is this 1,000-year period of observation that allowed the ancient Maya to learn all these things. Now you see why I think that this calendar really was created by these brilliant ancient people and not extraterrestrials.

The Day the Calendar Starts Is Linked to Izapa

The current cycle of the Long Count calendar started about 5,125 years ago. The beginning date was August 11, 3114 BC, Gregorian. On that day, the sun passed virtually directly over Izapa at high noon. There is almost no way for the sun to ever be higher in the sky but it was not quite perfect. My astronomy program shows me that it got to about 89 degrees, not a perfect 90 degrees and that the next day is just a bit closer to 90 degrees. Be that as it may, I still think that we can safely state that on the day the calendar started, the most important object in the sky basically rose almost all the way up to the highest and most supreme position in the sky. Is this just a coincidence? I don't think so. A Maya shaman told me that days when this happens are special and are considered to be a new beginning.

The latitude of the observer is what determines which days will have the sun travel directly overhead and you need to be within about 23 degrees of the equator for this to happen. If you are too far north or south, the sun will never be able to go directly over your head. Every year for Izapa, the sun goes overhead on May 1st and then again on August 12th. By picking August 11, 3114 BC, we see that the location of Izapa, Mexico, is integrated into the Long Count calendar. (Yes, I know; they missed by one day but within the constraints of their counting system, they came as close as they could. Or perhaps my software missed by one day; after all, I am asking it to project back in time about 5,125 years!)

I also want to point out something about the overall length of this remarkable calendar. As we know, the 1,000-year observational period allowed the ancient Maya to predict astronomical events 2,000 years into the future from the end of this observational period. But it also allowed them to determine astronomical events that happened 2,000 years before the beginning of this observational period. This gives us a total time span of approximately 5,000 years. This is why they were able to construct a calendar that had important astronomical events at both the beginning and the end of this long time span! And as we learned earlier, they intentionally picked the length of 5,125 years to underline their accurate understanding of the precessional cycle. Fantastic!

(For those of you who want to double check this astronomy, please note that astronomy programs refer to 3114 BC as -3113 so you need to enter a negative number for the year. You also need to use the Julian date of September 6th, not the Gregorian date of August 11th.)

We have just finished covering all the astronomy that I think is relevant to 2012. But many people are concerned with something a bit different called "the galactic alignment." For those of you who are curious, let's go into the reasons why I do not think that this is the reason that the Maya created the calendar.

The Galactic Alignment of John Major Jenkins

It is very important for me to point out that John Major Jenkins was the first person to link the restart date of the Maya calendar to an astronomical event driven by precession and he refers to this event as a "galactic alignment." John presented his ideas in 1995 in his book *The Center of Mayan Time*. John is well aware of the Maya's metaphor of the dark rift as the birth canal and he clearly sees the sun's presence in the dark rift as a rebirth. But there is something astronomers call the galactic equator that is very near the middle of the dark rift and this is what John focused on, perhaps because its exact location is precisely defined by astronomers. Strictly speaking, the dark rift is not even an astronomical object and because of this, astronomers pay little attention to it. It is the result of dust blocking the light traveling to us from the center region of the galaxy. This dust is not in our atmosphere; it is scattered throughout deep space between us and the center region of the galaxy. This creates the irregularly shaped dark section of the sky and as such, it does not have a precisely defined midline.

At the same time that John was solidifying his galactic alignment thesis, independent astronomers verified that the center of the disk of the sun would be almost exactly on the galactic equator at the precise moment of the winter solstice in the year 1998. John noticed that this alignment would occur repeatedly in the years around 1998 with only slight variation from one year to the next. Because of this, John decided to pick a range of 36 years that he centered around 1998 as the intended target of the Long Count calendar and he refers to these years as the "alignment zone" or "era-2012." During these years, the disk of the sun, as seen from earth, will be "touching" the galactic equator at the moment of the winter solstice. Let's take a look.

The Galactic Alignment of John Major Jenkins

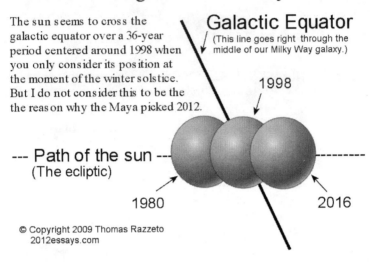

The sun seems to cross the galactic equator over a 36-year period centered around 1998 when you only consider its position at the moment of the winter solstice. But I do not consider this to be the the reason why the Maya picked 2012.

Galactic Equator
(This line goes right through the middle of our Milky Way galaxy.)

1998

--- Path of the sun ---
(The ecliptic)

1980   2016

© Copyright 2009 Thomas Razzeto
2012essays.com

Here is the way I like to explain it. At the moment of the winter solstice of each year, imagine that you take one picture of the sun and mark the position of the galactic equator. I call these pictures "time slices." If you play all these time slices like a movie, you will "see" the sun slowly move to the right across the galactic equator over this 36-year period. But notice that this collection of time slices is not something that you would actually see unfold in the sky above you. It is an unnatural way of "seeing" since you are only "watching" one moment per year. And what you are missing is the sun sweeping to the left for almost a complete circle throughout the year. So this is an artificially created scenario that has a solid foundation of truth but it can be easily misunderstood as the sun crossing the galactic equator by slowly moving to the right, over the course of several decades. Yet it ignores that the sun, as seen from earth, crosses the galactic equator each year in about 15 hours by moving to the left.

I have tremendous respect for John as a person, as a researcher and as a scholar. While his work linking the astronomy to the calendar was groundbreaking, I do not find this zone approach to be very satisfying. During my initial research, I seriously doubted that the Maya missed their intended target by 14 years, even though that is still a very small error, since there was so much precision being demonstrated by the creators of the calendar in other ways. I felt there must be something special about the exact day of the winter solstice of 2012 as seen from the specific location of the Maya. This is why I went further with my own original research and thinking and I came to the conclusion that *both* the triple rebirth of the sun *and* the special configuration of the sacred tree flying through the sky over the Maya on the restart date was the target of the calendar.

Both the restart date and the place where the calendar was created are important. We cannot ignore that this is a Maya calendar, not an Egyptian calendar or a global calendar. While the sun will be exactly in the middle of the dark rift during the day for the Maya, this event will happen during the night for people on the other side of the world. So the triple rebirth metaphor will not play out very well for those locations. But this is not a problem since the Maya were only concerned with what will be over them during that day, not what will be over others elsewhere, and, in my opinion, their message is for everyone anyway.

My view focuses on the center of the dark rift, not the galactic equator, although again, these two things are very close to each other. But more importantly, my view focuses on what will happen on that one day over the Maya, not what will happen over a long period of time as seen from the earth in general.

Also, please consider the fact that naked-eye skywatchers cannot detect the exact moment of the winter solstice and that no one can ever look up into the sky and see the galactic equator for it is as invisible as the equator of the earth. Both the earth's equator and the galactic equator are precisely located by scientists yet both are just imaginary reference lines. It is important to note that the Maya *could* look up into the sky at certain times and actually see the dark rift with their naked eye and this makes the astronomy useful for their metaphorical folklore. As I have repeatedly mentioned, I think this is an essential point in understanding 2012.

The idea that the Maya restart their calendar because of the combination of the triple rebirth of the sun and the unique configuration of the sacred tree is to the best of my knowledge, original with me. While other people are aware of the solar rebirths and the sacred tree in general, to my knowledge no one has connected them to 2012 like I have.

I want to be very clear. John was the first person to show that the astronomy and the calendar are related via precession and all his astronomy is factually correct. He is not wrong in any way; it's just that he is emphasizing a different aspect of the astronomy than I am. My focus is visual rather than abstract and much tighter in time. His approach leads to the years around 1998; my approach leads to the exact day of the winter solstice of 2012.

Chapter 4
Understanding the Mystical Aspects of 2012

The Wonder and Awe of the Miracle of Life Itself!

Did the Maya make the calendar simply because they could even though it was difficult? Was it a way of boasting or did they want to give us something, perhaps an invaluable treasure? The calendar clearly points to the rhythms of nature. What could they possibly want us to think or do in response to pondering the continuously reborn cycles of the day, the year and the great year? Their metaphor involves lovemaking and birth. Do they want us to contemplate the question of life itself? Do they want us to ponder the process of creation and the source of our very being? Do they want us to contemplate who we are at the deepest level?

We have all seen a newborn baby. While holding this new unique, precious person, many of us have been awestruck by the thought that this is indeed a miracle. And yet, how many of us have forgotten that all of us have never stopped being this miracle? When we look into the world around us, each person we see is a miracle and when we look into the mirror, we see a reflection of one of the most interesting examples of this miracle.

Perhaps the Maya are simply pointing to the rhythms of nature as a way of pointing to the miracle of life itself. The motion of nature *is* the motion of life! Without providing us with simple, obvious answers about what will happen to us in 2012, perhaps their intention is to inspire us into being in a state of wonder and awe about life as a profound mystery.

Life can be difficult. It takes hard work to maintain our existence on the physical plane. This daily burden can wear people down and even break their spirit. In addition to that, it seems that some people lose their sense of wonder when people explain things with science. Indeed, in today's world, many people believe that science has explained what life itself actually is. But has science really explained the mystery of life? Do you personally feel that you know what life really is? This not knowing is an essential aspect of the creation of this state of wonder and awe and this state is one of the treasures that is waiting for all those who dive deeply into this mystery.

2012 Is a Timeless Metaphor About Transformation and Rebirth

I think that the ancient Maya are trying to inspire us into being fully alive and they want us to recognize the opportunities that we always have in every moment, not just in 2012. For me, the sacred triple rebirth of the sun is a timeless metaphor about spiritual awakening. It is about what *can* happen *now*, not what *will* happen *then*. For me, 2012 is about holding each and every new moment in wonder and awe as the eternal now is continually born anew. For me, 2012 is about awakening to *our eternal awareness*.

Earlier, I mentioned that according to John Major Jenkins, the purest meaning of 2012 is found by examining the stone monuments and carvings in Izapa, Mexico, the birthplace of the Long Count calendar. This is the most immediate and primary evidence left by the Maya. In order for us to tune in to what the Maya are trying to tell us, let's take a closer look at what is found at this important site, which is located in the southern-most tip of Mexico in the state of Chiapas near Tapachula, just west of Guatemala.

Izapa, Mexico:
The Message of 2012 in Symbolic Form

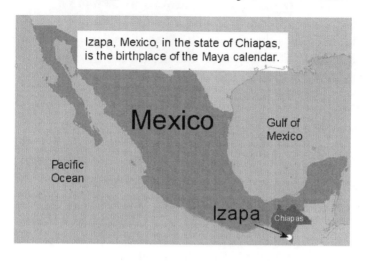

Izapa, Mexico, in the state of Chiapas, is the birthplace of the Maya calendar.

Mexico

Gulf of Mexico

Pacific Ocean

Izapa

Chiapas

In the sacred ceremonial site of Izapa, we find an important ball court that points to the location on the horizon where the sun rises on the day of the winter solstice. Many of you already know that the sun does not rise from the same place everyday. On the summer solstice, it rises from the northernmost point and on the winter solstice, it rises from the southernmost point. During the year, it moves back and forth between these two points. And for thousands of years, this ball court has pointed to the rising sun on the day of its yearly rebirth. Yes, once a year, this ball court simultaneously marks the daily rebirth and the solstice rebirth – a double rebirth of the sun.

The game that is played on this court adds even more to our understanding. John Major Jenkins explains that the long, narrow ball court is a symbol for the long, narrow band of stars in the Milky Way. Halfway down the ball court is a stone ring with a hole in the center, which serves as the goal. The stone ring represents the pregnant belly of the Sacred Goddess and the opening is the birth canal. The ball is the sun. The game is over when one team puts the ball through the center of the ring and we have the galactic rebirth of the sun. When we add the fact that the court is oriented towards the rising winter solstice sun, we see that the game is a metaphor for the triple rebirth of the sun in 2012. This is remarkably simple yet very rich in meaning. And let's continue just a little bit more. At one end of the ball court there is a throne, which allows the king to look down the long court at this special sunrise. On the front of this throne is a carving of a large round ball, symbolic of the sun in the birth canal coming forth into our world.

While there are other references to the astronomical triple rebirth metaphor in Izapa, I would like to give just one more example: the stone carving on Stela 11. Here we have the solar god emerging from the mouth of the Bufo toad with his arms and hands fully outstretched as if he is showing the full length of something. John Major Jenkins discusses this stela on page 67 of his 2009 book, *The 2012 Story*. John explains that the mouth of the toad is simultaneously the symbolic portal to the underworld and also the physical dark rift of the Milky Way. All things come into our physical world – "the world of the seen" – from "the unseen world" or "the underworld." The sun emerging from the underworld is a birth or a rebirth and this happens, as we of course know, when the sun is in the middle of the dark rift, the metaphorical birth canal, the portal where the spiritual world transforms itself into the physical world. But there is one more crucial point. This stela is facing the location on the horizon where the sun will rise on the day of the winter solstice. So this stela's image and position taken together clearly point to the triple rebirth of the sun in 2012. Again, this is fairly simple yet its full implications are profound.

(Due to copyright restrictions, I cannot show this stone carving here but you can search the internet for Stela 11 in Izapa to see it.)

John states that the deity's outstretched arms depict the fullness of an age (or the ending of an age, since it has reached its fullness). We of course know that many endings immediately lead to a new beginning and this new beginning is represented

60

by the triple rebirth of the sun. In my view, December 20, 2012 is the last day of the current Long Count calendar cycle and this of course means that December 21, 2012 is the first day of the new world age. *The calendar itself is reborn on the same day that the sun is reborn three times.* The deity is holding a cross in each hand and there is a circle on the crossing point of each cross. I think that this is referring to the sun on the crossing point of the sacred tree and that this marks the point in time when the precessional cycle is reborn.

So in Izapa, we see that astronomy and the theme of rebirth are repeatedly woven together. But what is this rebirth metaphor actually referring to? One of the most significant experiences that we can personally have is our own awakening to the spiritual nature of reality and it is easy to understand why many cultures have used the rebirth metaphor to point to this awakening. Is there any evidence that the Maya rebirth metaphor is pointing to something along these lines? Absolutely! It is now time for us to meet the most important clue found in the sacred site of Izapa: the Bufo toad.

The Bufo Toad and the Psychedelic Shamans

The Maya and their predecessors, the Olmec, were very interested in the Bufo toad for several reasons. One obvious reason was as a symbol of both transformation and rebirth. Not only does this creature transform itself from a small aquatic tadpole with a long tail and gills into an air-breathing amphibious toad over six inches long that can live a few decades, it sheds and eats its skin in a remarkable way. Approximately four times a year, the adult toad sheds its skin in one piece and it devours the skin as it is being shed. It might be said that the toad actually eats itself! The older outer form ceases to exist and yet it is absorbed and born anew as part of the toad's living body. Notice that this rebirth takes place without the death of the toad.

While all of this is important, there is another aspect of the toad that really brings it to the forefront and that is this: The Maya shamans used the psychoactive chemical excreted from the poison gland of this toad during their sacred rituals and in this way, each shaman had a direct, personal experience with what they called "the invisible world," "the unseen world," or "the underworld," and what we might call the spirit world. Please note that this chemical is obtained without killing the toad; some people refer to this process as "milking" the toad.

By the way, this psychoactive chemical is called 5-MeO-DMT and it has a somewhat different effect from other types of DMT from different sources, such as the vine used to make ayahuasca in Peru. Now, it is important that I clearly point out that this chemical is both illegal and risky. Remember, it comes from the poison gland of a large toad. Because of this, I strongly recommend meditation rather than psychedelics as a tool for self-discovery.

Please understand that these toads were so important that the shamans kept them as sacred pets. Since the toads live to be over thirty years old, many of them would outlive their owners. Thousands of skeletal remains of these toads have been found in mass graves and it almost seems as if the Maya and the Olmec created special burial sites specifically for these toads. One possible example of this is located in San Lorenzo and it dates back over 3,000 years, perhaps even 4,000 years. We should note that San Lorenzo is in the same general region as Izapa.

In the sacred site of Izapa, there is a stone carving, stela 6, which shows a shaman in a small canoe above the large open mouth of a Bufo toad. The canoe tells us that the shaman is on a journey and the open mouth, like a cave, is a metaphor for the entrance into the underworld. The gland of the Bufo toad that produces the psychoactive chemical is highlighted and in this way we clearly understand that this carving is a reference to the mind-altering trips that can lead to profound paranormal experiences.

Are these experiences the key to understanding the Maya transformation and rebirth metaphor? This is one of the most important questions in this book.

(Due to copyright restrictions, I cannot show this stone carving but you can search the internet for Stela 6 in Izapa to see it.)

While the subject of psychedelic shamanic rituals is often completely overlooked by the academic community, it is hard to dismiss the dual role these toads played for the Maya. Yes, of course they were symbols of transformation and rebirth because of their own physiological shift from tadpoles to toads and their frequent shedding of their skin. But the use of their mind-altering, paradigm-shifting chemical cannot be ignored. I think that the presence of the Bufo toad in the sacred ceremonial site of Izapa speaks volumes.

Stone statues of this toad were used as altars in this sacred site, which emphasizes that the toad was supremely important. It could not have been given a higher place of honor! If someone talks about 2012 and they don't talk about the Bufo toad, they are missing the most important clue of all!

(To see these altars, search the internet for Izapa stone toad altars.)

When you take into account the many references to transformation and rebirth in the Maya folklore, it starts to seem as if this concept is the Maya's favorite subject! Some of the scenes from this folklore are depicted in the stone carvings in Izapa. In my opinion, the evidence is very strong that this sacred site is primarily dedicated to the general idea of transformation and rebirth.

So we see that thousands of years ago, Maya shamans began to explore the full nature of reality via psychedelic rituals. Not only did they use 5-MeO-DMT, but they also used mushrooms and perhaps other chemicals as well. Some people will point out that the use of mind-altering chemicals will not help you find out anything at all about reality; it will only offer you hallucinations and misperceptions. This point is worthy of much discussion but for our purposes now, I will simply offer one point for contemplation. One experience that the shamans had was direct, personal interaction with spirit entities, often repeatedly with the same entities. If you were a shaman, could you imagine an entity from the other world greeting you with the following words:

"So you have decided yet again to visit me and still you wonder if I am really real. But I have a question for you. How do you know if your own world is not just happening in your head? How do you know if even you yourself are not just an illusion?"

While I do find those questions interesting, let's get back to our subject. The shamans could have also experienced astral projection. While this subject is fairly complex, it is very clear that in the opinion of those who do this, these experiences leave no doubt that a materialist worldview is woefully incomplete. Just this realization would be worthy of the rebirth metaphor yet I believe that the Maya shamans went much further than this. In this regard, let's see what an expert has to say about the effects of 5-MeO-DMT.

James Oroc, the author of *Tryptamine Palace*, stated in an interview that he had the experience of becoming "consciousness without identity." Since this is sometimes reported by others, it is extremely likely that some of the ancient Maya shamans also had this experience. In this way, it might be said that this experience can lead to a very deep awakening: the direct, personal discovery of the true, fundamental self – the One Awareness – which is much deeper than your personal self. Your normal personal self is what Oroc calls "identity." When you step back from that, you discover by direct experience that your true self is pure awareness. You know this because you *are* this! And you intuitively know that your pure awareness *is the One Awareness*.

64

What you think of as your personal consciousness is simply the One Awareness looking out through your own eyes! This is happening to you right now and it has been happening since you were born! What was created when you were born was not a new personal consciousness but a new viewpoint from which the One Awareness experiences the world! Your unconstructed self is the One Awareness; you are not the sentient being you have considered yourself to be all your life!

Now, not everyone who uses the chemicals from the toad awakens to this new understanding of the self, which is a core ingredient of what some people call "enlightenment." Yet the people that don't wake up to the true self often still wake up to the idea that there is a greater spiritual reality behind our ordinary physical reality. This is like waking up or being born into a spiritual world. This is why it can be said that the unfolding of this spiritual understanding has three births. Let's take a look at them.

Your Three Births

As we know, astronomically, the first rebirth is the daily sunrise. This is just the sun reappearing to us in our physical world. Likewise, with our physical birth, each of us also makes our appearance in our physical world. Both the astronomical rebirth and the biological birth are very common. One happens every day; the other happens to everyone. With this birth, some people take on a materialist view of reality. Since they haven't had any spiritual experiences, it makes perfect sense for them to see the world devoid of an unseen creator God or any spiritual aspects at all.

The next stage of development happens when the materialist has a personal experience that is beyond the explanation of their current understanding. This could easily come about in many ways, including the use of the chemicals from the Bufo toad. With this new understanding, the physical world is seen as a denser expression of a more energetic yet unseen spiritual world. The flow of this spiritual energy gives rise to what comes into existence in our physical world. As you know, these spiritual energies can often be felt by people who work on tuning in to them.

The final stage unfolds when you realize that you are pure awareness – the One Awareness – and that you are not the combination of your body, mind, soul, personality and personal consciousness. This re-identification as the One Awareness is the deepest shift that a human can go through. It is an experience of recognition. When you are born, you get an apparent identity. When you awaken to this understanding, you are reborn in the most profound way possible. Very few people have this happen to them yet I think that it happened to at least some of the ancient Maya shamans and that they intentionally created the triple rebirth metaphor and placed it on top of the astronomy of 2012 for the specific purpose of helping people understand these three stages of spiritual development.

So the three stages of your spiritual development are:

1) your start with a materialist's view
2) your shift into a spiritual understanding and
3) your awakening to your true self as the One Awareness.

In summary, we see that archaeological evidence shows that people in Mesoamerica over three thousand years ago began to track not only the stars and the planets but also precession. At that time, they also began to explore the full nature of reality via psychedelic rituals. The knowledge and wisdom from both of these things developed for over a thousand years and then came together in the creation of the Maya Long Count calendar.

Incidentally, just because an ancient culture had shamans does not mean that their philosophy of life was perfect or that their entire society functioned at a high level of consciousness. In this regard, please let me quickly talk about one of the most gruesome aspects of the Maya culture: human sacrifice. In this book, I am exploring the possibility that in the distant past, at least some of the Maya shamans were mystics and perhaps even fully enlightened beings. But even enlightened beings pass away and frequently those left behind do not fully understand the mystical view offered by these enlightened beings. As time passed, their message might have become misunderstood or even ignored completely. This might be how the Maya fell into the practice of human sacrifice.

What Is Not in the Sacred Site in Izapa

While I have known for years that the general region of Izapa was first designated by Maya expert Michael Coe and others as the birthplace of the Long Count calendar, I was recently quite surprised to read on page 63 of *The 2012 Story* that there are no Long Count dates recorded anywhere within the boundaries of the sacred site of Izapa itself! Why? If December 21, 2012 were so important, it would seem simple enough to at least put it on a stela that depicts the triple rebirth of the sun in 2012. With all the hype that 2012 is receiving today, you would think that this Long Count date was carved in stone 100 feet high at the very center of this important site, right?

But it is nowhere to be found! Think about that for a while!

Instead, throughout Izapa we see symbols pointing to the general concept of rebirth and specifically to the triple rebirth of the sun. This supports what I am saying: the time of 2012 and therefore the conditions of the world at the time of 2012 are not the focus. The concept of transformation and rebirth is the focus. The restart date is simply the time when the astronomical metaphor plays out in the sky. I like to explain it like this:

If I write a poem about *enlightenment* and *spiritual awakening* and use the sunrise as a metaphor since that is the event that fills our dark world with *light* and we are all *awakened* from our physical slumber, the poem should be taken as a reminder that spiritual awakening is possible in a general sense. It certainly should not be understood to say that we are all going to become enlightened during the next sunrise, although that may happen. If people focus on *the event of the sunrise* because they think enlightenment will happen *because of that event*, they will have missed the point of the poem. The same thing is true for the sacred triple rebirth of the sun in 2012. It is a reminder that we can all awaken to a new spiritual level at any moment. Focus on the *concept of spiritual rebirth*, not *the time of the event* that was used as the basis of the rebirth metaphor.

Very often the messages handed down from ancient people are metaphorical, not literal, since this can more easily point to timeless spiritual wisdom. In this way, I believe that the message of 2012 is timeless. If the focus of the calendar is timeless and everyone is looking for a literal event (other than the astronomy), they will miss the whole point of the calendar.

In my opinion, the conditions of the world are not at all the point of the message of 2012. That is just the outer form of creation and everything that has been created will change. The message of 2012 is much more profound. It is about something that never changes. It is about the fundamental principle of reality, as revealed by the psychedelic experiences of the shamans. I cannot stress this enough. The ancient Maya discovered the greater reality that lies behind our ordinary reality and most importantly, they discovered their true fundamental self, which was never born and will never die. So the questions of 2012 are: "Who am I?" and "What is the true nature of reality?" Asking what is going to happen to us in 2012 is much too superficial.

The message of 2012 is about the fundamental nature of reality, not about the conditions of our world at this time.

The distinction that I am making is subtle and often overlooked, and since I want to be clear, here's a recap. The calendar points to one specific day. On that day, an astronomical event will unfold that will not cause anything unusual to happen to us. Yet that astronomical event is being used as the foundation of the metaphor of the triple rebirth of the sun. But the metaphor is only a symbol. Our reward comes when we discover what the rebirth metaphor is pointing to: a personal spiritual rebirth that can happen at any time.

When I first started researching 2012, I was tightly focused on the astronomy of the calendar and I'd like to take just a minute to explain a key point regarding that astronomy. In order to understand the astronomy of 2012, you only need to understand the motion of the earth. To produce the triple rebirth of the sun, the earth goes through three different motions: it spins on its axis once a day, it travels around the sun once a year, and it wobbles on its axis once every 26,000 years. The triple rebirth of the sun does not come about due to anything happening to the sun itself; it comes about due to the shifts in our relationship with the sun.

These three motions have been repeating for billions of years and there is no reason to expect that the astronomy will suddenly cause something unusual to happen to us in 2012.

I think that the Maya also knew this, and this is why it made perfect sense for them to use the astronomy as a timeless rebirth metaphor rather than as an alarm for danger or a time marker for world peace. In my opinion, this is an extremely important point that is completely misunderstood by many people.

You Are Eternal Like the Sun

I think that the Maya personified the sun because you are like the sun. From our perspective on earth, the sun is born when it rises and it dies when it sets. But from a perspective out in space, we see that the sun is alway shining. It is never born and it never dies; it is essentially eternal.

Similarly, from the perspective of our ordinary reality, it appears as if we are born and then later, we die. Is there a different perspective, a larger framework, that reveals that our true self, our fundamental unconstructed self, is indeed also without birth or death? Are we also eternal? Yes, absolutely!

I think that the Maya used a timeless metaphor on purpose to help you understand that you, too, exist beyond the boundaries of time itself. You exist as pure awareness, and time, space, energy and matter are all created within you. While many people have heard that the world is an illusion, very few people have heard someone say that they themselves are also an illusion. Yes, indeed! Your personal self is an illusion!

When people hear me say that their personal self is an illusion, they think that I have just said that their personal self is not real. But that is not what I mean. Remember, all illusions have a reality to them; it's just that they trick you into believing a false idea. In this case, there are at least two false ideas. One is that your personal consciousness arises out of the biochemistry of your body and brain and the second is that each personal consciousness is separate from all the others. Those ideas are completely false.

Yes, it seems like you are a separate, mortal, sentient being but you are not a being of any kind; you are sentience itself! This is the formless divine essence! This is what the gnostics focus on when they say, "Know yourself and you shall know God!"

Many of us have heard it said, "You are not a human being having a spiritual experience, you are a spiritual being having a human experience." Of course it is wise not to put the cart before the horse but perhaps you now see why I say that you are neither the cart nor the horse! You are the awareness that holds both the human being and the spiritual being!

Your eternal nature is looking out of your eyes right now, shining forever, just like the sun!

Chapter 5
Mystical Spirituality

Life Is an Adventure of Discovery

Perhaps life has no goal, either in the outer conditions of your personal world or even in your inner experiential states. Life is an adventure of discovery and as such, nothing is known ahead of time; it all comes forth spontaneously. In an adventure of discovery, there is of course the next step on the journey but that step is not taken by a true explorer until they have thoroughly experience what is there for them in that moment. *Drink it all in completely!* Inner peace and liberation from dissatisfaction are side effects of the posture that you take and the purity with which you hold without clutching everything that comes to you. Think about that. Peace, liberation and even spiritual growth are not the goal; they are side effects.

Some people think that when I talk about the pure experience of the moment that I am talking about removing all distractions. For example, I was walking on the beach during a spectacular sunset and I saw a couple walking side by side. Instead of holding hands, they were holding their cell phones, which had completely captured both of them. If their intention was to enjoy being together while walking on the beach with all the magnificent colors, the smell of the salt air, the sound of the crashing waves and the cries of the seabirds, they had become distracted and completely missed all of that. Obviously, I recommend not becoming distracted but what I am really talking about is removing the extra layer of false beliefs about yourself and the world that pollute your experience in every moment.

We have all heard the phrase: be here now. Yet somehow it seems to me that when people hear this phrase, what they actually hear is: be here now ... with all your stuff. No, it's simpler than that. You can leave all your stuff behind!

This section of the book will examine the balance between the passive and the active. I will present ideas that will assist you in holding a posture that will allow you to glow with the flow and yet also allow you to actively step forward into the world and consciously express your creativity. Let's start off by talking about peace.

The Path of Peace:
The Shift from Competition to Cooperation

Many teachers talk about healthy and unhealthy fears. A healthy fear will arise when there is a clear and present danger and your body and mind will react immediately in a natural way. This type of fear does not occur very often and therefore it will not lead to a state of constant fear. But unhealthy fears, which are imagined fears about the future, can lead to a constant state of fear. The body and mind were never intended to handle this situation and this can create problems for many individuals.

Things can become even more problematic if these unhealthy fears spread throughout society and we end up creating a fear-based society. When I first heard that phrase several decades ago, I did not fully understand its depth. Now I believe that I see things more clearly. Every aspect of the very foundation of our society is literally built upon false beliefs that are permeated with fear. This makes it much more likely that new individuals joining our society will adopt these beliefs. So this circle feeds back on itself and it can become quite a challenge for anyone to break out of it.

All of this has resulted in a fragmenting of our society to the point where everyone is simply watching out for themselves. We have lost the fundamental supporting cooperation of the group and replaced it with the struggle of the individual. In this way, everything has become a competition. We start to compete at the youngest age in order to win friends and get better grades in school. Then the competition continues for the best colleges, jobs and spouses. This competition is not harmonious and when it grows and spreads throughout the many nations of the world, this disharmony comes forth in war.

Competition can only lead to war because each step along the path is a little war.

Yet when we cooperate with one another by putting into action the values of caring and sharing, we enhance the well-being of all the individuals in the group. This has the power to bring us something that humankind has seldom seen. This has the power to bring us peace. The gift of peace is a gift we give both to ourselves as individuals and to ourselves as a group.

To understand how we can bring forth our birth as conscious creators of peace, we need to understand how the process of creation actually works. It is helpful to note that we don't need to become fully enlightened; we only need to see the relationship between cooperation and peace on both the personal level and on a larger scale. So now let's take a look at inner creation, outer creation and the circle of unconscious creation.

The Process of Creation:
Inner Creation and Outer Creation

Perhaps you have noticed that the thoughts and attitude that you bring to each and every moment determine your emotional experience of that moment. If you bring the thoughts and attitude of acceptance, you will experience the emotions of wholeness, peace and joy. If you bring the thoughts and attitude of rejection, you will experience the emotions of separation, frustration, fear and pain.

I call this inner creation. Inner creation instantly creates any human emotion from pure agony to total ecstasy. Each and every human being can experience all of these emotions. And, while we all can create these emotions consciously by choosing our thoughts and attitude, most people unconsciously create them with habitual thoughts and reactions.

You have probably heard people say, "I can't help it. That is just the way I feel." But when we dig deeper into why we feel an emotion, we notice that emotions arise from the dynamic interaction of perception, ideas and beliefs. If we change an idea or belief, we may experience a completely different emotion. This may inspire us to reconsider an underlying belief. In this way, we can tune our worldview by weeding out false beliefs. I won't go into this any more right now but you can see that we are not really helpless victims of our emotions. We can take conscious responsibility for the emotions that we experience and this will help us be reborn as conscious creators of peace.

While the emotions of inner creation come forth instantly, outer creation is a process that unfolds through time. It is the attraction of people, things and events into your life. These are all outer conditions. I don't believe that all outer conditions are available to everyone. In other words, we all have different menus, so to speak. One person may be able to become world famous and yet another person, seemly similar in every respect, may achieve only minor success, as measured by society, even though she or he works wisely and diligently with a positive attitude.

Nonetheless, we all have numerous choices and certainly many of them will be in harmony with our personal growth. It is somewhat like a farmer. He may choose to plant several crops and put forth all the appropriate effort but the farmer has no guarantee that anything will grow at all. Normally, most of his crops do grow but on rare occasion, a severe drought may stifle the crops completely leaving nothing but dirt.

Your thoughts, attitude, perceptions and beliefs, some of which are just habitual thoughts, instantly create your emotions, and all of those things attract the conditions of the outer world as time unfolds. The outer reflects the inner and it all begins with a thought.

When you are aware of the process, you can consciously choose. When you are unaware of the process, you just react habitually. If you see a problem in the world, you will try to fix it by manipulating the conditions of the outer world directly without going to the true source of the problem, which is the underlying false belief. This is a fairly involved subject and I won't go into it more now but if you want to learn more, I recommend a book by Jane Roberts called *The Nature of Personal Reality – A Seth Book*. It offers both concepts and exercises that can help you recreate the foundation from which you build your life.

In this way, I encourage people to consciously work toward a better future. Yes, like the farmer, plant new seeds! But don't harshly blame yourself if it does not turn out the way you had envisioned. Just accept it as "what is" and move forward.

Seek what you love; love what you find!

This is the posture that I am talking about. Have an idea about what you want to bring forth in the future but don't be attached at all to how it will unfold.

I have seen spiritual communities become very focused on "manifesting" whatever you want. It seems that the most popular books and DVDs are often about making the material world match your endless desires. The fundamental belief that is being catered to is this: I can be happy when I get what I want and this is how I get what I want. But there is a deeper understanding that offers all of us much more than the conditional happiness that we experience when we get what we want. Sure, that is always fun but what if you were fully satisfied with everything just the way that it is now? This is liberation!

And still this leads us to some very interesting questions: To what degree do we control our own life? How much do we do and how much just happens to us? Is it wise for us to work with the understanding that we have some control but not total control? Many of you know the simple but profound Serenity Prayer:

"God, grant me the serenity to accept the things I cannot change, the courage to change the things I can, and the wisdom to know the difference."

Within the common understanding of our personal self, I think that this is an excellent posture to hold. No one can know exactly what the future will bring. Many situations are like a baseball batter trying to hit the ball. Just do your best and accept what happens. I know that other teachers say, "Don't *try* to do something; *just do it!*" but this doesn't seem to hold up in my experience. Yes, not knowing what is going to happen to us is an essential aspect of what makes us feel alive and I think that it is one of the most beautiful things about being human because it creates a sense of adventure, a sense of discovery. And when it comes to your unknown future, remember, God may have a wonderful surprise for you at any moment! If we always knew what was coming, there would never be any surprises.

What would life be like without surprises?

And have you considered shifting the idea that things happen *to* you to the idea that everything is happening *for* you? This is a very bold thought. Seth would add that this happens in accord with your conscious beliefs. Can you try this idea on for a while? Most people will dismiss it immediately without any real inquiry. Will you be like most people or will you ponder it? Will you discover how it is true for you?

The Circle of Unconscious Creation

More than anything else, I would like to bring forth more caring, kindness and peace. With every new moment, I have the opportunity to place peace and love into my heart and then nurture these seeds through time.

Yet if I hold an emotionally charged judgment in my mind and heart in the form of frustration, anger or hatred regarding the current conditions of the world, those emotions and thoughts will, in time, bring forth new disharmonious outer expressions such as tension between other people and perhaps even violence or war.

Since I may also emotionally reject the new outer conditions that arise, I may find myself stuck in a circle of unconscious creation, a circle of disharmony. This is basically what is happening all throughout the world today in every aspect of society.

Consider the social tension about war. Often, some people support a particular war while other people speak out against it. If you could look into the hearts and minds of all these people, what would you find? I suspect that you might find plenty of judgment and anger, and perhaps even hatred and fear in both groups. The military generals may hate the enemy and be angry at the peace demonstrators while the demonstrators may hate the war and be angry at the generals. And both groups may fear a terrible outcome if they don't get their way. But all this judgment, anger, hatred and fear will not bring us peace; it just attracts dramatic outer conditions that match the turmoil felt within.

In order to reap a peaceful future, seeds of peace need to be planted in the present. A world without war is built by people with peace in their hearts.

It is important to note that the process of creation is a private, personal experience under the direct control of each individual. *Your personal experience does not depend on what others do.* Yes, there are dynamic energetic connections between people but the important point is that each individual has control over their own thoughts. These thoughts are crucial in the creation of your emotional experience, and all of this attracts certain outer conditions as time unfolds.

Yes, we reap what we sow and we can shift into personal peace by letting go of judgment and by practicing kindness. This, in time, will bring forth new peaceful outer conditions. Try it yourself and see what arises!

Handling Challenging Emotions – Be Fully Alive!

And yet, sometimes very difficult situations arise, such as the passing of a loved one. The idea is to respond to whatever arises in a natural and genuine way as free from judgmental thought as possible. In other words, I don't necessarily think that the goal is to be happy all the time; I think that the "goal" is to be fully alive in the moment!

Animals experience fear in a natural and genuine way but only at the appropriate time. They also experience joy and sadness. Similarly, we can experience these emotions as part of the natural flow of life without being chained to the pains of the past or in constant fear of what the future may bring.

Don't think, "If I was more spiritual, I would not feel so sad." When you find yourself feeling deep sorrow, don't be afraid of it or judge it; dive into it head first! Be fully alive! You may surprise yourself by discovering that while it can be a very uncomfortable emotion, after it passes, your true essence, your awareness, is not damaged at all.

Life is a package deal. Birth eventually leads to death.

Be alive with such fullness that even if the emotion is deep sorrow, as this emotion courses through your being resonating with every cell, there is still an underlying healthy sense of tremendous vitality that comes from your fearless thrusting of yourself into the moment with the thought,

"I am fully alive! I have the power to feel all human emotions and these emotions will never harm my true self!"

And what is this thing we call "badness?" It is just part of the subjective experience that we create when we add harsh, judgmental thoughts to our perceptions. But is it really bad to be sad? Is the *quality* of badness in the *essence* of sadness? It is the constant running around in such a way so as to avoid what we normally call "bad" and seek out what we call "good" that brings us an unnatural life. Unawakened people focus on getting everything they want. Mystics know that they can safely experience and embrace whatever shows up!

Embrace the whole of life and you will embrace the whole of God!

Conditional Happiness

Now of course, many of us do experience periods of time when we are pretty happy and perhaps even ecstatic. Life is certainly not all suffering, even if we're not enlightened. When we have things the way we want, we give ourselves permission and we choose to "push the happiness button," which is on the inside. I call this "conditional happiness" since we choose it only when our worldly conditions are acceptable.

Yet these moments of joy may also present a problem. How can we make them last? How can we hold on to them? If we don't make the conditions last, our happiness also disappears. Yet we all know that everything will change – sometimes quickly, sometimes slowly but nevertheless all things do change.

Parents see their little children, whom they want to hold in their arms forever, grow up and fly away into the world. How much courage does it take to love them fully while they are home when you know that one day they'll be gone? What does it mean to fully love something and yet be perfectly okay emotionally when it goes away? Is this even possible? Well, for many people, this can present quite a challenge.

A Deep Divine Bliss Can Hold All Human Emotions

The present moment is like a gift, a "present." And it's a magical gift that becomes what you make it. The attitude that you hold as you unwrap the present determines whether your experience is joyful or sorrowful. This may seem like an over-simplification so I think that it is important for us to dig deeper into this subject.

I've heard Buddhist teachers speak about the end of suffering but perhaps the phrase "the end of dissatisfaction" will be more helpful to you. In other words, physical and emotional pain will always be a part of the human experience. That's just the way that it is. Yet your attitude still holds tremendous power. With it, you create your experience of the moment and attract the conditions of the future. As you know, if you bring an attitude of rejection, you will experience dissatisfaction, to be sure!

And to add to what I said earlier, if you bring an attitude of acceptance, you may even experience a fundamental divine bliss that can comfortably hold human sadness within it. In other words, you may experience human sadness in its purest form without judgment. In this way, you are sad but you are not suffering. Think about that for a while. Ponder it deeply.

It is like a divine mother who knows how to stay centered in her divine bliss while holding her crying child in her loving arms. She would never push away her dear child or judge it as undesirable in any way. You can be like the combination of the two of them with your divine bliss holding your human sadness.

With this understanding, you awaken to the fact that while you may indeed be able to accept more and more of the world in a way that brings genuine human happiness, it is not necessary for you to be happy all the time since all human emotions can be held by the larger framework of this deep divine bliss. You will be able to experience human sadness from a point of view that allows you to remain free from suffering any dissatisfaction.

Now let's consider an example where the shift away from dissatisfaction might be readily available. This is not a very profound example at all and yet it often leads to strong emotions. Here it is. If someone cuts you off in traffic, our society almost demands angry thoughts, a hateful attitude and even a rude hand gesture. With these choices, negative emotions often surge through every cell of your body. Similarly, our society forbids you to emotionally embrace moments that are connected with injury, sickness, death or tragedy and requires you to judge these things as "bad" and unacceptable.

Do you really want to follow society's dictates when they only lead to suffering?

You do not need to be happy *that* the person cut you off in traffic but you can still be happy *in spite of it*. Are there examples in your life where something used to bother you a lot but now when they happen, it's no big deal? Can you learn to joyfully accept more and more of the world exactly as it is?

But much more significantly, if you genuinely have a broken heart, can you step back into your deep divine bliss and simply hold that broken heart? The first time I did this was a very powerful experience. As the experience unfolded, I witnessed a mountain of deep sorrow dissolve right back into my awareness. It was then that I realized that all human emotions were created out of my awareness and that this is where they would return.

Indeed, all experiences arise from the One Awareness; the mystery of creation is that the formless Awareness gives birth to all that has form and this is what creates all of our experiences.

By the way, if you will, ponder this: If something really should be different, wouldn't it actually be different? If God did not want something to exist, it would not exist. If God were truly against something happening, it would not happen. For whatever reason, God allows everything that comes forth.

I have heard some people say that in order to awaken spiritually, you need to have your heart win its battle with your mind. But please consider this. Your mind holds ideas; your heart holds emotions. Your mind can hold true ideas or false ideas. Your heart can hold hatred and fear or love and peace. At the deepest level, only false ideas can prevent you from being at peace. There is no battle where your heart must triumph over your mind. Instead of a battle, think of it as the pursuit of truth and peace. Work towards tuning your mind into the truth and your heart into love and peace.

The Mystic's View: God and Creation Are the Same Reality in Different Forms

Once you start to think about the concept of transformation and rebirth, you notice that it occurs in many things and in many ways. And most importantly, we notice that certain shifts in our own attitude and actions can actually bring forth our own rebirth. Is this what the Maya are trying to inspire us to do? Becoming fully alive by completely embracing the world exactly as it is and awakening to the process of conscious creation are two important rebirths but again, our deepest and most fundamental rebirth occurs when we awaken to the divine nature of all of reality and discover our true identity.

I think that the ancient Maya not only wanted us to contemplate who we are at the deepest level, I think that they wanted us to actually find the answer. Yet the answer to a question this deep is not appreciated as much when it is simply given to you. Furthermore, the answer is often resisted or even rejected if it is pushed upon you from an apparent authority. Answers to questions this deep really begin to mean something only if you find them for yourself. Yet it can be significantly helpful if someone points you in a fruitful direction and perhaps even assists in triggering your own mystical experience for the core of this wisdom is not available through logic alone. If that were true, many people would have "gotten it" by now.

In order to help explain this fundamental rebirth, which is an awakening to what you have always been, I like to start with a simple metaphor about water and ice. As we all know, water and ice are both made out of the same thing: hydrogen and oxygen, H_2O. Yet even though water and ice have the same essence, they are very different in their form. Scientists call this different "states" and these states give rise to a completely different set of properties, a completely different set of qualities. As we all know, ice is cold and solid and water is warm and fluid. Let's expand the water analogy to include water vapor, an invisible, formless gas. That might be like God the Source. We cannot see God the Source since the Source has no form.

And yet, if the Source wanted to create a spiritual world, a world of angels, for example, the Source could choose to lower the temperature, the vibration, of the H_2O that is the divine essence of all things and part of the Source would change from an invisible, formless gas into liquid water that now has some form to it. This spiritual world could include spiritual beings with a sense of self and freewill. They would not appear to be the same as the Creator and yet they would be made out of the same divine essence. Beings in this world would have their senses tuned to that realm so they would be able to perceive and experience everything in it.

And if the Source wanted to create a physical world, to create our universe, the Source could choose to lower the temperature even more and part of the Source would then take on many magnificent new forms. Similar to before, this physical realm could include physical beings with a sense of self and freewill. And again, they would not appear to be the same as the Creator and yet they would be made out of the same divine essence. These beings would, of course, have their senses tuned to the physical world, which would allow them to perceive and experience everything in it but they would generally be blind to everything in the spiritual world.

Now, most analogies fall short in some respect and the water analogy is no exception since it is missing the transcendent aspect of the Creator, yet I like it in many ways. But is it true?

Is everything we see actually the eternal divine essence in physical form? Does God intentionally hide from us in order to invigorate the play we are all in and charge it with emotion? Can we see God hiding in the world just like we might see a secret assistant to the magician? Is this divine essence the core of the wonder and awe that we feel when we admire the beauty of nature and each other?

In this mystical view, it's important to note that while we are all made out of the same essence, we are not all the same. From the viewpoint of our common daily life, we each have our own unique "personal self," precious and special but not in an egotistical sense. We each have our own body, we each have our own personality and we each have our own thoughts and feelings and so forth.

Imagine for a moment that the eternal divine essence was wood. You can make a chair out of wood, you can make a house out of wood and you can make a boat out of wood. While all these things are made out of the same thing, they are each very different in their form. If you are out on the ocean, the boat will serve you best and if a storm is coming, the shelter of the house will benefit you the most.

So don't make the mistake of saying, "Well, since everything is divine in its essence, everything is the same in its form and I will not make any distinctions between people or things." *Some people and things may mix with you in a more harmonious way so choose wisely and enrich your life!*

And yet, with this mystical understanding, you will perceive every act as a divine act, every thought as a divine thought and every emotion as a divine emotion. Indeed, every thing will be seen as the divine in physical form.

You are not simply connected to God; every aspect of your being is divine! God is not just deep within you at a special place; you are divine at every level!

In the formless state of the eternal divine essence as source, there is only oneness, wholeness, and yet in that extraordinary state (which is not really a state) nothing is manifest. It is here where we all dissolve into oneness. Yet in our physical world, the wholeness of the divine essence is presented to us in a magical way that makes it appear as separate, non-divine pieces. This is the illusion of the world. But just because the world is an illusion does not make it not real for it is real as an experience. Remember, an illusion has some reality to it, yet it tricks you into believing a false idea.

Throw away the false idea but don't throw away the reality of the world for it is the divine essence wearing a disguise!

Duality and Oneness

Here we see the importance of always remembering the essence while appreciating the significance of the form. In this way, we can understand the paradox of duality. The two sides of the same coin have different shapes yet they literally share the same essence. So it is with all examples of duality.

I sometimes hear people talk about an ideal world where we all move beyond duality. Yet all of our experiences arise because of duality. Without duality, there would be nothing to experience. And we do all move into non-duality every night during deep dreamless sleep since there are no experiences at that time. (Take some time to think about that.) Furthermore, some people move into non-duality while they are wide-awake during meditation. What I think people mean when they talk about a world without duality is a world without emotionally charged judgments. This is liberation from binding likes and dislikes. If this were true for everyone, then we would have a world full of liberated beings living happily together in duality. Wouldn't that be something!

The Chinese symbol of yin-yang beautifully weaves together the concepts of duality and oneness. Of course we have all heard it said, "We are all one." While I certainly think that this is true, many people find it to be quite confusing so let's dig into the subject of oneness a little bit further.

Some people intuitively know that we are all one. I want to be very precise here. They have the experience of perceiving an intuitive feeling that we are all one. Yet they are not *directly perceiving* the One Awareness that makes us all one since the One Awareness has no aspects that can be perceived or experienced. *It has the capacity to see yet it cannot be seen; it has the capacity to hear, yet it cannot be heard and so forth.*

While the intuitive feeling of oneness is true, people often try to validate it by finding pure oneness in the world of form. They usually show how we are all alike or how we are all connected and they offer this as examples of how we are all one. Yet there is no pure oneness in form. In form, you can find closeness, intimacy, compassion, empathy and so forth. Again, your body and mine are different and the same can be said for our minds, our personalities and our souls. This is why I say that in form, we are many. What you can see in the world of form is that light and dark are complements of each other. They go together in a

86

way that points to fullness and wholeness. But even together, they are not pure oneness; they are constructed from the essence of the One Awareness.

Some people point out that oneness is not something that can be understood by the mind alone and this is true. They go on to say that oneness is something that you can only know through experience. This needs some clarification. All relational experiences occur when the observer perceives something with qualities and aspects. You can perceive a table, a nighttime dream, an emotion, a thought, an intuitive feeling and so forth. But you cannot perceive the One Awareness. You cannot experience this through a relationship of observer and object. And yet you can know this because you *are* this. You know you are pure awareness *directly* by being "it" and to go even further, you know through an intuitive experience that your pure awareness *is* the One Awareness we all share. These two steps together show us that we are all one.

Yet instead of just saying, "We are all one," I like to say:

"In form we are many; in essence we are one."

Or:

"In form we are human; in essence we are divine."

If you ask me, these are two of the most important sentences in this book.

Your True Self: Your Pure Awareness

Paradoxically, when the formless divine essence becomes manifest, it presents itself with form, with an appearance. It is this physical appearance that our scientists focus on and in many ways they do an excellent job. Yet few of them have any interest in the awareness that actually sees the world. Of course, some of them study the personality or the physiology of perception but that is not the awareness. This wondrous awareness is what you actually are.

Your awareness is the buried treasure hidden within the personal self; it is the kingdom of heaven within you.

Your "personal self" may be your body, your mind, your personality and even your soul but your true self – your fundamental, unconstructed self – is the awareness that perceives everything that you experience.

Honest scientific research into subjects such as out-of-body experiences, near-death experiences and past lives all point to a nature of reality that is profoundly different from what is commonly believed.

Our amazing awareness itself is not even visible to us and yet we know it exists because we experience the world through it!

Mainstream science holds that the material qualities and chemical processes of the body and the brain give rise to our awareness. In other words, in this view, the material world is the foundation of reality and awareness is an aspect that comes forth from it in order for all sentient beings to be alive and aware. In this view, our awareness depends upon the body.

But what if it's the other way around? What if awareness is the foundation – the building block – of everything?

Our common daily experience shows us that there is a link between our body and our awareness. In addition to that, past-life research shows us that while there is a relationship between our awareness and our body, the body is not the source of our awareness. The true nature of the self is beyond the body. Out-of-body experiences also support this understanding. And the same thing holds true for our soul. We are not our soul and that is not where we find our eternal self. We are our awareness, which exists before our soul and holds our soul.

Of course we all see that our bodies were born into the world; this is beyond dispute. But isn't it also true that both our bodies and the world are born into our awareness? Our awareness is primordial; it comes before everything and it plays host to all the phenomena that we experience. And a mystic intuitively understands that the Divine Awareness that looks out of your eyes is the same awareness that looks out of his or her eyes. What is seen is different but the Awareness is the same. There is only One Awareness, that of the eternal divine essence.

Imagine you are the One Awareness and that you are in a room with many windows. When you look into the world through one of these windows, you see a certain view. In this way, you have all the experiences of one particular person. You see what they see, feel what they feel, think what they think, remember what they remember, dream what they dream and so forth. When you look through another window, you have all the experiences of another person. But you are still the same awareness. If you experienced all of these experiences all at once, it would be just a mass of color and noise. We might say that our individuated personal consciousness is what allows the One Awareness to experience the world without it being a jumbled mess. But is there really an individuated personal consciousness? What if personal consciousness does not really exist as such? What if it is really just the One Awareness looking through a window? The One Awareness can never be cut into pieces and it does not have any parts. It does not give birth to new, separate "awarenesses."

Even so, it still can be helpful to use the phrase "personal consciousness" when we talk about ourselves and our experiences. In this way, perhaps a good definition of personal consciousness would be that which holds both our personal experiences and our personal beliefs about the world and about our own self. It is this collection of beliefs that changes and evolves as our personal consciousness moves into a higher state. In this process, we are actually just removing false beliefs and . integrating new truthful understandings. Your personal consciousness simply holds these new understandings. That which is held does change but that which holds it all simply exists as the unchanging capacity to hold.

Another way to look at the "One in All" or the "All in All" is to imagine that there are billions of hand puppets on earth, all apparently different people and animals. But when we see beyond the illusion of separate sentient beings and correctly perceive the full picture of what is really taking place, we see that there is only one sentient puppet master with billions of hands animating all of life and, most importantly, experiencing all of life through all those different viewpoints. What a miracle! What a mystery!

Obviously, this does not mean that you should cast off all concern for the personal self. Your personal self provides tremendous functionality. While it is not you in the most fundamental way, your kind and caring stewardship of your personal self will offer the one divine Self a very rich human experience. In a more complete sense, referring back to the water and ice analogy, as the ice, you are the entire physical world; as the water, you are everything in the spiritual world and as the invisible formless water vapor, you are the source of all creation, which is your unconstructed fundamental self.

In the fullest sense, you are the totality of reality and you are where it all comes from!

Once the mystic sees the whole world as the eternal divine essence in physical form, everything becomes precious and sacred. Deep compassion for all people, animals and even plants springs forth automatically. Yet even before we realize the mystic's view, while we are still tricked by the illusion of separateness, we really can see that we are at least all connected to one another like a big family and we can use that realization as motivation to shift into cooperation.

Don't Overlook the Changes That Are Happening to You

It does not matter if the Maya foresaw the trying times that we now find ourselves in. Nor does it matter if they predicted world peace. That is not important at all. What is important now is that we correctly assess our situation and do our best to solve these challenging problems. No matter what the Maya were thinking, it seems to me that humankind is indeed dealing with tremendous challenges on a global scale. Many of our activities are not only fundamentally unsustainable, they appear to be at or close to the breaking point right now. In fact, some people argue that in certain areas, we are already past the breaking point and dominoes are now falling into other dominoes and a collapse is already unfolding.

Yes indeed, there are many serious problems facing all of humanity. In some ways, things may look completely dire. But consider the circumstances of an unhatched chick. Imagine that it is you who is trapped in the tiny space of the egg. Obviously, since you are the only one in the egg, it is perfectly natural for you to focus on yourself. Each day, some of the food is eaten and no new food is ever delivered. As the supply of food is about to be exhausted, you think that you are surely doomed.

But you have overlooked something. You have overlooked the changes that have been happening to you. You are now strong enough to break out of the shell and into a whole new world that you never could have even imagined – a world filled with family, friends, plenty of room to roam, blue sky and lots of food! You have entered a world where you are now naturally doing your part by simply being yourself. While this is radically different in form from when you were in the shell, we should not overlook the fact that it is also true that you were naturally doing your part then, too.

Is it possible that a similar dramatic shift is waiting for all those who have the wisdom and courage to embrace an attitude of cooperation and break through the shell by putting into action the values of caring and sharing without fear of what might arise? Is this the rebirth that is awaiting you as the next stage of your natural development? Is this the rebirth that is awaiting all of humanity? Will we break out of our shell of competition and be reborn as conscious creators of a peaceful world, naturally united in cooperation? And is this development automatic, just like a seed growing into a mighty tree? Of course this *is* possible and it is always exciting to watch what unfolds in the world!

We all have magnificent opportunities before us and one of the most fruitful shifts awaiting us now, both as individuals and as a group, is the shift from competition to cooperation. When we all sing out in unison, "*One for all and all for one!*" we will all live in peace.

By the way, we have all heard it said, "As above, so below." But what does this phase actually mean and is there any truth to it? Notice that it does not necessarily mean that what is in the sky creates or controls what is on earth. My opinion is that at its core, this phrase is stating that all of creation and everything that happens is a reflection of the mind of God, the source of everything. The power and will of God is making all of this come forth in a natural and divine way.

It is the divine that is above it all!

Mystical Optimism

When it comes to 2012, should you be an optimist and see the glass as half full and the future as bright *because you think that it will match what you desire?* Or is the glass half empty and big trouble just around the corner?

Both the optimist and the pessimist make a fundamental mistake when they believe that the very essence of the future can be either good or bad. Again, the experience of goodness or badness is not intrinsic to outer conditions; they are part of the subjective experience that arises from the judgmental thoughts that we add to the mix.

For me, reframing the glass as half-full rather than half-empty seems like a shallow, emotional manipulation since it does not change what I actually have or change what I am aware of. This doesn't interest me as much as discovering, for example, that the glass can be filled again from a nearby faucet or stream. Now that makes a real difference!

But certainly some situations do not reframe so easily into a desirable scenario. What if you are in a desert and that half-glass of water is all you have for a two-day hike to safety? This is where the view of the mystic comes in. As I mentioned before, a mystic realizes that God and creation are the same reality in different forms. Ice does not just come *from* the water; it *is* the water!

The mystic's view offers us the most profound reframing available. With this mystical understanding, you can joyously embrace an optimistic attitude like never before. I call this "mystical optimism" and with it you can fully embrace whatever you have before you right now and also whatever comes tomorrow because you see it all as fully divine and you know that whatever form it takes, it can never harm your true self, your divine awareness!

So put mystical optimism and the values of caring and sharing into action at every moment. Work towards a more desirable future in any way that appeals to you. Do so joyfully with hope in your heart and without fear of what will arise. While it is not our nature to know what will arise next on the physical plane, we can *glow with the flow* and *dance with the divine* in a way that allows us to experience the deepest richness that life has to offer no matter what shows up!

Yes, embrace the whole of life and you will embrace the whole of God!

~~~ The End ~~~

Thanks for reading my book! I hope that you have found it to be both inspirational and transformational. While I do enjoy it when people tell me that they find my work inspirational, that is not why I write. Inspiration without proper guidance leads nowhere. I write to awaken you to the divine nature of all of reality and to help you discover the true identity of your fundamental self. May you experience this now as your ultimate rebirth!

Have a magical and mystical day!

Thomas Razzeto
February 28, 2012

End Notes:

All my 2012 essays and a few videos are at:

www.2012essays.com

You can find all of my mystical spirituality essays at:

www.infinitelymystical.com.

Most of my work is completely free to read from my website and some essays have free Adobe PDF files that make them easy to print on 8.5 by 11 sheets of paper.

Please note: I used to give *Mystical 2012* away in its entirety from my website as a free PDF. But because of the terms for selling ebooks online, I am no longer allowed to do so. But don't worry; the work I sell will never contain any "secret teachings" only available to those who have paid money! My free essays will always contain all the important information.

Suggested reading:

The Nature of Personal Reality - A Seth Book by Jane Roberts

The Afterlife Experiments by Dr. Gary Schwartz

Children's Past Lives and Return from Heaven, by Carol Bowman

Life Before Life by Dr. Jim Tucker

Robert Monroe's three books on the out-of-body experience: *Journeys Out of the Body, Far Journey* and *Ultimate Journey*

Dr. Jeff Long's Near Death Experience Research Foundation:

http://www.nderf.org

Made in the USA
Charleston, SC
07 September 2012